BEYOND THE FINITE

BEYOND THE FINITE

The Sublime in Art and Science

EDITED BY

ROALD HOFFMANN AND IAIN BOYD WHYTE

Oxford University Press, Inc., publishes works that further
Oxford University's objective of excellence
in research, scholarship, and education.

Oxford New York
Auckland Cape Town Dar es Salaam Hong Kong Karachi
Kuala Lumpur Madrid Melbourne Mexico City Nairobi
New Delhi Shanghai Taipei Toronto

With offices in
Argentina Austria Brazil Chile Czech Republic France Greece
Guatemala Hungary Italy Japan Poland Portugal Singapore
South Korea Switzerland Thailand Turkey Ukraine Vietnam

Copyright © 2011 by Oxford University Press

Published by Oxford University Press, Inc.
198 Madison Avenue, New York, New York 10016
www.oup.com

Oxford is a registered trademark of Oxford University Press

All rights reserved. No part of this publication may be reproduced,
stored in a retrieval system, or transmitted, in any form or by any means,
electronic, mechanical, photocopying, recording, or otherwise,
without the prior permission of Oxford University Press.

Library of Congress Cataloging-in-Publication Data

Beyond the finite : the sublime in art and science / edited by Roald Hoffmann and
Iain Boyd Whyte.
  p. cm.
 ISBN 978-0-19-973769-7 (hardcover : alk. paper) 1. Science—Aesthetics.
2. Sublime, The, in art. I. Whyte, Iain Boyd, 1947–
 Q175.32.A47B49 2010
 500—dc22   2009054357

1 3 5 7 9 8 6 4 2
Printed in the United States of America
on acid-free paper

For striking the initial sparks, our sincere thanks to Felice Frankel, a distinguished artist, an advocate of better seeing in science through photography, and the indefatigable organizer of the Image and Meaning conferences. We are grateful to Suhrkamp Verlag and its editors Hans-Joachim Simm, Regina Oehler, and Sabine Landes for publishing this volume in German as part of its Edition Unseld. At Oxford University Press we found a welcome home, and have received good advice from Jeremy Lewis, Lisa Stallings, Hallie Stebbins, and Theresa Stockton. We also appreciate the expert help of Claudia Heide in obtaining good quality images and permissions for their use. Roald Hoffmann is grateful to Jennifer Cleland and Catherine Kempf for research and editorial assistance.

Editors' Preface

There are many excellent books on the sublime written by philosophers, aestheticians, and art and cultural historians.[1] This volume represents a first attempt to extend the discussion of the sublime into the realm of the natural scientist. The project originated in a conversation among a chemist, a behavioral neurologist, an art historian, and an architectural historian. The occasion was a conference entitled "Image & Meaning 2," held at the Getty Center in Los Angeles in June 2005. The conversation took place in a session of this conference entitled The Sublime in Art and Science, and the ambition, which finds a broader rehearsal in this book, was to search for areas of commonality in which a topic widely discussed in art and aesthetic theory could be shared with and opened up to the discourses of the sciences. Perhaps the exchange might offer science some insights as it discovers and creates the new.

The chosen vehicle—the sublime—is broad enough in its many definitions to stimulate new thinking both in the arts and in the sciences. The sublime has meant many things over its long history as it has been applied to the emotional impact of that which is beyond beautiful. Philosophers have contemplated the anxiety that the sublime provokes and the resulting joy of self-assertion that it offers in the face of an uncontrollable world. In the words of Jean-François Lyotard, the sublime is "the pleasure that reason should exceed all presentation, the pain that imagination or sensibility should not be equal to the concept."[2] The breadth and indeterminacy of the term are central to the project. Rather than address the sublime head-on as a category seeking definition, this volume uses it as a catalyst to provoke responses from a group of distinguished scientists and cultural historians.

In this context, the sublime is not offered either as a veiled religiosity or as a mandate for nihilism. Rather, it is seen as a means of defying conceptual rules and, in the process, relating insights that were formerly unknown to each other. As the philosopher Kirk Pillow has suggested, "Sublime reflection can provide ... a model for a kind of interpretative response to the uncanny Other 'outside' our conceptual grasp. It thereby advances our sense-making pursuits even while eschewing unified conceptual determination."[3]

The best of science also makes claims to the sublime, for in science as well as in art, each day brings the entirely new, the extreme, and the unrepresentable. How does one depict negative mass, for example, or the propagating, contagious misfolding of a protein? Can one capture emergent phenomena as they emerge? Science is continually faced with describing that which is beyond previous experience and common sense. That, too, might be a definition of the sublime.

And scientists have begun to discern the physiological and genetic basis of our senses of the beautiful and the sublime. Contemporary neurophysiology is charting the landscape of our emotions, while biology addresses the transformation of sensory impulses into aesthetic judgments.

Editors' Preface

One might not have thought an aesthetic category would be a fruitful meeting ground of art and science. This book, by nine courageous scholars, proves that it is. Their explorations range far—from the Hubble Telescope images to David Bohm's quantum romanticism, from Kant and Burke to the "downward spiraling infinity" of the twenty-first-century sublime, through a denial of utility to the category itself, to *Little Nemo* and the affective foundations of the sublime. The sublime brings nine free-thinking authors together, offering a uniquely powerful portal into both art and science, and into as yet uncharted spaces between.

Notes

1. A list of major recent monographs on the sublime includes Paul Crowther, *The Kantian Sublime: From Morality to Art* (Oxford: Clarendon Press, 1989); Christine Preiss (ed.), *Das Erhabene: Zwischen Grenzerfahrung und Größenwahn* (Weinheim: VCH Acta Humaniora, 1989); Andrew Ashfield and Peter de Bolla (eds.), *The Sublime: A Reader in British Eighteenth-Century Aesthetic Theory* (Cambridge: Cambridge University Press, 1996); Kirk Pillow, *Sublime Understanding: Aesthetic Reflection in Kant and Hegel* (Cambridge, MA: MIT Press, 2000); James Kirwan, *Sublimity* (New York: Routledge, 2005); Gene Ray, *Terror and the Sublime in Art and Critical Theory* (New York: Palgrave Macmillan, 2005); and Philip Shaw, *The Sublime* (Abingdon, UK: Routledge, 2006), which also has an extensive bibliography on the sublime.

2. Jean-François Lyotard, *The Postmodern Condition: A Report on Knowledge*, trans. Geoff Bennington and Brian Massumi (Minneapolis: University of Minnesota Press, 1984), p. 81.

3. Pillow, *Sublime Understanding*, p. 2.

Contents

Contributors | xiii

1 The Sublime: An Introduction | 3
   Iain Boyd Whyte

2 Affective Foundations of Creativity, Language, Music, and Mental Life: In Search of the Biology of the Soul | 21
   Jaak Panksepp

3 Still Deeper: The Nonconscious Sublime; or, The Art and Science of Submergence | 43
   Barbara Maria Stafford

4 Pretty Sublime | 57
   Elizabeth A. Kessler

5 Against the Sublime | 75
   James Elkins

6 Neuroscience and the Sublime in Art and Science | 91
   John Onians

7 Quantum Romanticism: The Aesthetics of the Sublime in David Bohm's Philosophy of Physics | 106
   Ian Greig

8 Disobedient Machines: Animation and Autonomy | 128
Scott Bukatman

9 On the Sublime in Science | 149
Roald Hoffmann

Index | 165

Contributors

Scott Bukatman is an Associate Professor in the Film and Media Studies Program in the Department of Art and Art History at Stanford University. He holds a Ph.D. in Cinema Studies from New York University and is the author of three books: *Terminal Identity: The Virtual Subject in Postmodern Science Fiction*, one of the earliest book-length studies of cyberculture; a monograph on *Blade Runner* commissioned by the British Film Institute; and a collection of essays, *Matters of Gravity: Special Effects and Supermen in the 20th Century*. His writing highlights the ways in which popular media (film, comics) and genres (science fiction, musicals, superhero narratives) mediate between new technologies and human perceptual and bodily experience. His latest book project, *The Poetics of Slumberland*, expands on the theme of disobedient machines in comics, cartoons, and *My Fair Lady*.

James Elkins is E. C. Chadbourne Professor in the Department of Art History, Theory, and Criticism, School of the Art Institute of Chicago. He writes on art and nonart images; his recent books include *On the Strange Place of Religion in Contemporary Art*; *Visual Studies: A Skeptical Introduction*; *What Happened to Art Criticism?*;

and *Master Narratives and Their Discontents*. He has edited three book series, including *The Art Seminar* (conversations on different subjects in art theory) and *The Stone Summer Theory Seminars* (stonesummertheoryinstitute.org).

Ian Greig is an artist and academic whose 2003 Ph.D. thesis, "The Aesthetics of the Sublime in 20th Century Physics," explored the aesthetic dimensions of physics and found a new location for Kant's theory of the sublime. Between exhibitions of his paintings, Ian has presented on this theme at various conferences exploring the confluence of art, science, and philosophy. He is currently the Postgraduate Lecturer at the National Art School, Sydney, Australia.

Roald Hoffmann was born in 1937 in Złoczów, Poland. Having survived the war, he came to the United States in 1949, where he is now the Frank H. T. Rhodes Professor of Humane Letters, Emeritus at Cornell University. In chemistry he has taught his colleagues how to think about electrons influencing structure and reactivity, and he has won most of the honors of his profession, including the 1981 Nobel Prize in Chemistry (with Kenichi Fukui). Hoffmann is also a writer of poetry, essays, nonfiction, and plays, carving out his own land between poetry, philosophy, and science.

Elizabeth A. Kessler is a visiting assistant professor at Ursinus College in Collegeville, Pennsylvania. She completed her Ph.D. at the University of Chicago in 2006, and she is currently working on a book on the aesthetics of Hubble Space Telescope images. She has been awarded fellowships by the National Air and Space Museum, Smithsonian Institution, and the 2006–2007 Sawyer Seminar on Visualizing Knowledge, Stanford University. Her work has been published in *Studies in the History and*

Philosophy of Science, in *Nuncius*, and in *Hubble: Imaging Space and Time*.

John Onians is Professor Emeritus at the University of East Anglia, Norwich, where he established and directed the World Art Research Programme. He has taught and lectured at many institutions in Australia, China, France, Germany, India, the Netherlands, the United States, and elsewhere and was the first Director of Research and Academic Programs at the Clark Art Institute in Williamstown, Massachusetts. He has held fellowships at the Warburg Institute, the Center for Advanced Study in the Visual Arts in Washington, D.C., the Getty Research Institute, the Wissenschaftskolleg in Berlin, and the University of Otago, New Zealand. He was the founding editor of the journal *Art History* (1978–1988) and edited the first *Atlas of World Art*. Among his books are *Bearers of Meaning: The Classical Orders in Antiquity, the Middle Ages and the Renaissance*; *Classical Art and the Cultures of Greece and Rome*; *Art, Culture and Nature: From Art History to World Art Studies*; and *Neuroarthistory: From Aristotle and Pliny to Baxandall and Zeki*. He is currently working on two books exploring ways in which neuroscience can help in the study of art, the first concentrating on Europe and the second embracing the whole world.

Jaak Panksepp (born in Tartu, Estonia, in 1943) is Baily Endowed Professor of Animal Well-Being Science at the Veterinary College of Washington State University and Distinguished Professor of Psychobiology, Emeritus, at Bowling Green State University. His research is devoted to the analysis of neuroanatomical and neurochemical mechanisms of emotional behaviors, with a focus on understanding how separation responses, social bonding, social play, fear, anticipatory processes, and drug craving are organized in the brain, especially with reference to psychiatric disorders.

He has written more than 400 scientific papers in the area and has edited eight books, most recently *The Textbook of Biological Psychiatry*. His monograph *Affective Neuroscience: The Foundations of Human and Animal Emotions* helped establish a new field seeking to understand the deep neuroevolutionary nature of mammalian emotions as they relate to basic sources of human values and the nature of emotional disorders in humans and animals.

Barbara Maria Stafford, born in Vienna, studied at Northwestern University, the Sorbonne, the Warburg Institute, and the University of Chicago, where she has taught for 20 years and for the last decade has held a University Chair. An avowed imagist, her writing focuses on the history and theory of imaging and visualization modalities from the early modern to the digital era. Her books, in various ways, reveal the deep intersections connecting the arts, sciences, and optical technologies to one another: geography, geology, and mineralogy (featured in *Voyage into Substance*); anatomy and the life sciences (*Body Criticism*); neuroscience and cognitive science (*Echo Objects*). She also writes historically grounded manifestos on the vital significance of the visual and sensory arts to general education as well as to society at large (*Artful Science; Good Looking*).

Iain Boyd Whyte is Professor of Architectural History at the University of Edinburgh and Director of VARIE (Visual Arts Research Institute Edinburgh). He has published extensively on architectural modernism in Germany, Austria, and the Netherlands and on post-1945 urbanism. Beyond architecture, he has written on twentieth-century German art and on Anglo-German literary relations. He is editor of the Getty Foundation–funded e-journal *Art in Translation*. A former fellow of the Alexander von Humboldt-Stiftung and a Getty Scholar, he was cocurator of the Council of Europe exhibition *Art and Power*, shown in London, Barcelona, and Berlin in 1996/1997. He has served as a Trustee of the National Galleries of Scotland, is a Fellow of the Royal Society of Edinburgh, and Chair of RIHA (the International Association of Research Institutes in the History of Art).

BEYOND THE FINITE

1

The Sublime

An Introduction

IAIN BOYD WHYTE

The test of a construct that has true value is its ability to retain its core yet take on diverse and different guises over the centuries, each of them expanding both the potency of the construct and our understanding of it. So it has been with the sublime. Dismissed in the heroic years of high modernism as a passé and febrile embrace of Kantian philosophy by the romantic imagination, the sublime enjoyed an enormous revival of interest during the postmodern 1980s and 1990s. With the certainties of the modernist project held up to critical reexamination, the sublime offered a vehicle with which to question the dominant view of human agency on which the modern economic and political order had been established. Dismissing as reductive and one-dimensional the modernist conception of the human condition as rational, progressive, and benign, the postmodern critique found in the sublime a device for exploring more profound and complex layers of meaning: the heroic, the mysterious, and the numinous.

The principal focus of this interest was the Kantian sublime. Toward the end of the 1980s, the aesthetician Paul Crowther noted a new confidence in defining the sublime, which had opened the category to discussions that extended far beyond the traditional parameters: "While philosophers in the analytic tradition of philosophy have found new significance in Kant's treatment of beauty and art, philosophers from other traditions and, indeed, writers in a host of other disciplines have asserted the cultural centrality of the sublime—and, in particular, Kant's version of it."[1] At that time, the sublime was highly fashionable in such fields as poststructuralist linguistics, literary studies, psychoanalysis, and psychology. It strongest impact, however, was to be found in the realm of art theory and criticism, particularly as applied to the high modernist and postmodernist avant-garde.

In the mid-eighteenth century, the sublime was seen to exist in great and awful objects. As Edmund Burke explained in his celebrated exegesis *A Philosophical Enquiry into the Origin of Our Ideas of the Sublime and Beautiful* (1757):

> Whatever is fitted in any sort to excite the ideas of pain, and danger, that is to say, whatever is in any sort terrible, or is conversant about terrible objects, or operates in a manner analogous to terror, is a source of the *sublime*; that is, it is productive of the strongest emotion which the mind is capable of feeling. . . . When danger and pain press too nearly, they are incapable of any delight, and are simply terrible; but at certain distances, and with certain modifications, they may be, and they are delightful, as we every day experience.[2]

And it was through language that the sublime object—be it craggy cliff, charnel house, or graveyard—could be related to our fears of pain, death, and destruction, where the objective could be tied to the emotional.

For Immanuel Kant, in contrast, writing at the end of the eighteenth century, the sublime exists not in an object that has tangible form, contours, and dimension. This is the key difference between the beautiful and the sublime. As Kant explains: "The beautiful in nature has to do with the form of the object, which consists in the boundary [*Begrenzung*]. The sublime, on the other hand, is to be found in a formless [*formlosen*] object, insofar as in it or by occasion of it *boundlessness* [*Unbegrenztheit*] is represented, and yet its totality is also present to thought."[3] In other words, we can know the ocean through our senses and dip our toes into the briny, but the vastness of the ocean is an idea that cannot be an object of sense experience, because it lacks contours and boundaries. Rather than in the things of nature, the sublime is to be found only in our own perceptions, when our inability to estimate the magnitude of things makes us aware of a supersensible faculty within us. This is the paradox of the sublime. The gap between tangible, empirical objects, on one hand, and the world of the supersensible, on the other, is absolute and unbridgeable. Yet only through our failure to represent the supersensible can we have a presentiment of its existence. In this very ambivalence lies the power of the sublime. As Slavoj Žižek explains: "This is also why an object evoking in us the feeling of Sublimity gives us simultaneously pleasure and displeasure: it gives us displeasure because of its inadequacy to the Thing-Idea [the supersensible Idea], but precisely through this inadequacy it gives us pleasure by indicating the true, incomparable greatness of the Thing, surpassing every possible phenomenal, empirical experience."[4] Even though our rational capacity to measure, assess, and comprehend is challenged by the sublime object, we take pleasure and consolation from the recognition that our supersensible capacities are not similarly constrained.

Mediated by an initial displeasure, therefore, at the inadequacy of our powers of reason to comprehend the object of our contemplation, we experience a state of delight at the triumph of our supersensible faculties. Provided they are viewed from a secure position, the most violent, powerful, and boundless manifestations of nature—Kant points to the hurricane and to the great cataract of water tumbling down from enormous heights—can engender feelings of joy: "We readily call these objects sublime, because they raise the forces of the soul above the height of vulgar commonplace, and discover within us a power of resistance of quite another kind, which gives us courage to be able to measure ourselves against the seeming omnipotence of nature."[5] This delight in our supersensible faculties can be prompted both by the magnitude of the object and by its quality—in short, its scale and its power—which Kant treats separately under the headings of the Mathematical Sublime and the Dynamic Sublime.

Kant is ambiguous in the *Critique of Judgement* on the relationship between the sublime and the work of art fashioned by human hands. On one hand, he specifically excludes as possible stimuli of sublime awe such things as buildings, monuments, statues, and the like, where a human end determines the form as well as the magnitude of the object. This does not preclude, however, the attempt to make artistic representations of those natural phenomena that have the authentic power to stimulate a sublime response. In this context, text fares better than visual art in Kant's estimation of sublime potential, with the suggestion that "even the presentation of the sublime, so far as it belongs to fine art, may be brought into union with beauty in a *tragedy in verse*, a *didactic poem*, or an *oratorio*, and in this combination fine art is even more artistic."[6] For Kant, however, the supersensible delight that might accrue from such works is related to their proximity with moral ideas. This may be seen to confirm the proposition that while the sublime is of dubious value as an aesthetic concept, it can stand indirectly for moral awareness.

Kant's intuition that dramatic text rather than visual art is a more potent vehicle for the construction of the man-made sublime proved prescient. While such tragedies as Friedrich Schiller's *Die Räuber* or Aleksandr Pushkin's *Boris Godunov* left few standing on the stage and no one in the audience unmoved by their relentless inquiries into the nature of evil, the sublime in painting tended toward kitsch. There are, of course, wonderful exceptions, such as J. M. W. Turner's alpine landscapes or his fantastic watercolor of the eruption of Vesuvius in 1817. (See Fig. 1)

Equally convincing is Caspar David Friedrich's *Chalk Cliffs on Rügen* (1818/1819), in which diminutive figures in the foreground peer tremulously down into the abyss formed by the chalk cliffs and spiky outcrops that plunge down into the Baltic on the island of Rügen. (See Fig. 2)

Indeed, the sequence of the three figures reading from left to right might be seen to exemplify the Kantian sublime, with the woman on the left pointing down in initial horror at the incomprehensible magnitude of the drop below. To her right, a male companion is momentarily incapacitated as his rational imagination fails to rise up to the vertiginous challenge before him. Farther to the right, however, a third figure leans nonchalantly against a tree, which provides a safe vantage point on the wondrous spectacle, as fear is transformed into joy in the experience of the sublime.[7] As Friedrich himself observed, we cannot escape from thoughts of death and transience: "To live eternally once, we must often surrender to death."[8] Through the intercession of the work of art, physical fear can be converted into pleasing astonishment. Generally speaking, however, the sublime impulse in nineteenth-century landscape painting led to bland sentiment: the sunsets of Frederic Edwin Church, the waterfalls and mountainous vistas of Albert Bierstadt, or John Martin's apocalyptic landscapes, "in which cataclysms are rendered on huge canvases with painstaking attention to the details of rending granite and screaming women."[9] The result is more ridiculous than sublime, endorsing the researches of the nineteenth-century German aesthetician Friedrich Theodor Vischer into the affinities and differences that exist between the beautiful, the sublime, and the comic.[10]

Partially escaping from these quasi-religious catacombs, however, the realm of the sublime expanded in the nineteenth century beyond the excesses of nature to embrace almost any area of human experience marked by great wealth or power. With the emergence of industrial production and urban concentration in the nineteenth century, however, the inventions of man rather than nature offered a new focus for sublime contemplation. As Crowther has noted in his study of the Kantian sublime: "The structures of capitalism and the conflicts it engenders provide immediate and inescapable images that overwhelm our perceptual or imaginative powers, yet make the scope of rational comprehension or human artifice and contrivance all the more vivid."[11] In the nineteenth century, industrial production, the speed and power of steam technology, and the burgeoning metropolis or industrial city stimulated the sensations of awe, terror, and exaltation previously associated with such natural phenomena as cliffs, waterfalls, and deserts. In the context of the Victorian cities in Britain, Nicholas Taylor lists among the sublime delights of the new century "the haranguing of the Evangelical preacher; the ecstasy of the Anglo-Catholic Mass; the scientific wonders of panoramas and exhibition halls; the traveller's thrill in catching trains and climbing mountains; the capitalist's pride in the hum of mass production and hubbub of the market."[12] The city of brick and stone, driven by the limitless technological power of steam and iron, with its vast and ever-expanding scale and its brutal contrasts of splendor and deprivation, replaced the menacing mountains, crags, and cliffs of the eighteenth century. The conquest of the Alps and the conquest of the industrial city demanded similar qualities and provoked parallel aesthetic responses. To the boundlessly large, one must add the minutely and incomprehensibly small, particularly when that which is invisible to unaided sight—the bacillus or virus, for example—has terrifying powers of destruction.

As the instrument that makes possible the victory of reason over nature or chaos, the sublime has always carried with it extra-aesthetic dimensions, both social and political. While Kant's "mathematical" sublime refers to the faculty of cognition, the "dynamical" sublime points to the realm of human ambition and desire. David Hume had already made this connection in his *Treatise of Human Nature*, in which he notes that, "in collecting our force to overcome the opposition, we invigorate the soul, and give it an elevation with which otherwise it would never have been acquainted."[13] Writing in the mid-nineteenth century, the German aesthetician Friedrich Theodor Vischer offered a more muscular account of the same insight, which proposed a distinctly positive understanding of the sublime: "We feel ourselves elevated because we identify ourselves with the powers of nature, ascribing their vast impact to ourselves, because our fantasy rests on the wings of the storm as we roar into the heights and wander into the depths of infinity. Thus we ourselves expand into a boundless natural power."[14]

With the sublime response so closely linked to perceptions of power and achievement, it is little wonder that the guardians of totalitarian aesthetics looked benignly on this aesthetic mechanism. Indeed, as Gary Shapiro has argued, "an exclusive poetics of the sublime can lend itself all too easily to irrationalist, fascist politics."[15] This nexus of sublime vision and concrete political ambition is entirely consistent with the logic of the sublime. A telling example from the early years of the National Socialist regime in Germany can be found in Martin Heidegger's exegesis of the Friedrich Hölderlin poems, written around 1800, in which the poet wrestles with mortal man's attempts to portray the workings of the gods. In "Wie wenn am Feiertag," for example, Hölderlin offers this dynamically sublime vision of the workings of poetry:

> Yet, fellow poets, us it behoves to stand
> Bare-headed beneath God's thunderstorms,
> To grasp the Father's rays, no less, with our own two hands
> And, wrapping in song the heavenly gift,
> To offer it to the people.[16]

The conclusions that Heidegger draws from this half-stanza are entirely compatible with the National Socialist ideology: "Thunder and lightning are the language of the gods, and the one whose purpose it is to bear this language without equivocation and place it in the being [*Dasein*] of the people is the poet."[17]

Although he precluded man-made works from the imperium of the sublime, Kant, as already noted, admitted that the works of man might, nevertheless, produce a sensation of delight. He surmised, for example, that when a visitor enters St. Peter's in Rome, the resulting emotions would lead to a calming sense of reassurance: "A feeling comes home to him of the inadequacy of his imagination for presenting the idea of a whole within which that imagination attains its maximum, and, in its fruitless efforts to extend this limit, recoils upon itself, but in doing so succumbs to an emotional delight."[18] In the context of National Socialist architecture, a similarly pseudosublime response—feeding off the tension between what is perceptually overwhelming yet still known to be artifice—would have awaited the visitor to Albert Speer's projected "Große Halle," a gigantomanic domed structure set as the terminal feature of an axis that would have cut through Berlin like a surgeon's scalpel.

The Sublime

Albert Speer's "Große Halle," model photograph, 1940, Private Collection, London

Courtesy of Lutz Becker

By its very nature, the sublime moves from an immodest overestimation of our human powers at one extreme to despair and nihilism at the other. The horrors of the Holocaust undoubtedly turned thoughts back to the latter, suggesting a world in which there was no meaningful reality beyond the finite.

A particularly powerful response to the nihilist position came in the visual arts from the New York–based color field painters, who, in the words of Barnett Newman, were "reasserting man's natural desire for the exalted, for a concern with our relationship to the absolute emotions."[19] Shunning the conventional props of history, myth, memory, or association, Newman and his contemporary Mark Rothko held that transcendent sublimity was the only defensible subject matter of painting. Preoccupied in the mid-1940s by Jewish myths of the Creation, Newman produced his large-scale "zip" paintings, in which broad, monochrome fields are divided by thin vertical lines in contrasting colors. Their titles, such as *Vir Heroicus Sublimis* (1950–1951) or *Adam* (1951–1952) confirm the transcendent ambitions of these works and the desire to present the unpresentable. These ambitions were underlined in Newman's writings from the period, and in particular by the essay "The Sublime Is Now," first published in 1948.

Newman's article was a key text for postmodern theories of the sublime, as formulated by Jean-François Lyotard in *La condition postmoderne: rapport sur le savoir* (1979). According to Lyotard,

> The sublime ... takes place ... when the imagination fails to present an object which might, if only in principle, come to match a concept.... We can conceive the infinitely great, the infinitely powerful, but every presentation of an object destined to "make visible" this absolute greatness or power appears to us painfully inadequate. Those are Ideas of which no presentation is possible. I shall call modern the art which devotes its "little technical expertise" [*son "petit technique"*], as Diderot used to say, to present the fact that the unpresentable exists.[20]

While Newman pointed explicitly to Burke as his principle source on the sublime, Lyotard and his contemporaries were more drawn to the Kantian, a priori sublime, grounded on the interrogation of reason and its limits.

A precondition for the postmodern critique of production in the visual arts was the conviction that the work of art emerged from and addressed not only the eye but also the intellect. In the twentieth-century context, however, the essentially Kantian belief that the function of the artwork is to stimulate in us the sense of a supersensible ability was not a postmodernist insight, but rather one of the key presuppositions of early abstraction. Paul Klee, for example, once noted in a poem that "I am not to be comprehended in *this* world."[21] Similarly, the claim to present the nonvisible was already entirely explicit in the founding manifesto of abstraction, Wassily Kandinsky's *Über das Geistige in der Kunst*, first published in 1912. In a language indebted to the rhetoric of the sublime, Kandinsky announces the end of the old order of representational art as it is challenged by abstraction and the dawn of a new era in which the resonance between the artwork and the soul forms the foundation of a new, spiritual revolution.[22] His description of the color white is symptomatic of his claims for the power of abstraction to satisfy our inchoate inner needs:

> This world is too far above us for its harmony to touch our souls. A great silence, like an impenetrable wall, shrouds its life from our understanding. White, therefore, has this harmony of silence, which works upon us negatively, like many pauses in music that break temporarily the melody. It is not a dead silence, but one pregnant with possibilities.[23]

The embrace of the spiritual realm by whiteness preoccupied Kandinsky in the early years of heroic modernism and was taken up again and embellished by the postmodern theorists at the end of the century.

The world changed radically, however, between Kandinsky's early-century optimism and the reinvestigation of the sublime in the late twentieth century. Writing in the 1960s, Karsten Harries proposed a binary model, in which the positive end of the sublime axis points to the infinite and to a numinous reality that transcends the palpable limits of man. This was accompanied by a negative sublime, predicated on the belief that the only way of confronting and combating despair lies within ourselves: "The only transcendence revealed is that of man himself. The finite is negated only to liberate the subject. The negative sublime is the epiphany of freedom."[24] This position, which was exemplified in the heroic optimism of high, rationalist modernism—particularly in architecture and urbanism—waned dramatically in the 1970s and 1980s. In consequence, the ambition to master the supersensible through the assertion of reason faltered in the postmodern sublime, which focused on the immanent rather than on the transcendent, on the objects of the world and on our powers of imagination rather than on the sacred and the mystical. The sublime experience, thus understood, no longer points to the supersensible that exists beyond reason, but rather to "that [quality] *within* representation which nonetheless *exceeds* the possibility of representation."[25]

The impossibility of both representation and rationalist appropriation made the sublime an obvious vehicle with which to approach terror, for as Lyotard has noted, the only possible politics of the sublime is terror. The philosophical engagement with the Holocaust, with the bombing of the German and Japanese cities, or the genocides in Cambodia and Rwanda was further sharpened by the events of September 11, 2001, which generated images in live time that exercised a morbid fascination on all who saw them in their countless repetitions. Television viewers around the world were forced by these images to employ faculties of mind that transcended every standard of sense and destroyed at a stroke the metaphors of the conquest of gravity proclaimed in two icons of twentieth-century technology, the skyscraper and the airplane.

Terrible catastrophes had, in the past, played a part in rethinking the sublime. The Lisbon earthquake of 1755 exercised an enormous influence on Immanuel Kant, who studied every available account of the catastrophe before publishing a slim volume on the theme in 1756.[26] The knowledge of this natural disaster, which killed a quarter of the city's 250,000 inhabitants, lurks insistently behind Kant's subsequent investigations of the sublime.

The attack on the Twin Towers of the World Trade Center has had a similar impact and has already generated new studies on the aesthetics of terror. The description of the site as "Ground Zero," with its reference—possibly guilt laden—to Hiroshima[27] and to the *Stunde Null* of Germany in 1945, indicates an abyss vastly more terrifying than the Baltic cliffs depicted by Friedrich. The presence in every television set of these images and our visceral reaction to them shift the argument away from the Kantian discussion on disinterested aesthetic judgment and the limits of reason, and back toward the eighteenth-century British discourse on the sublime. This earlier model is predicated much more vigorously on fear and terror, stresses the interdependency of aesthetic judgment and ethical conduct, and uses aesthetic response to interrogate how we, as subjects, make sense of our experience. How do we respond to an unimaginable event like the collapse of the Twin Towers? While the event itself, subsequently analyzed in minute detail, is not beyond comprehension, the images themselves are incomprehensible in their terror and are truly sublime.

Burke, as noted above, explicitly links the terrible and the sublime:

> Most of the ideas which are capable of making a powerful impression on the mind, whether simply of Pain or Pleasure, or of the

modifications of those, may be reduced very nearly to those two heads, *self-preservation* and *society*; to the ends of one or the other of which all our passions are calculated to answer. . . . The passions therefore which are conversant about the preservation of the individual, turn chiefly on *pain* and *danger*, and they are the most powerful of all the passions.[28]

The terror engendered when human existence is threatened by forces beyond its control or comprehension is the dynamo, therefore, of the sublime aesthetic. A precondition for this powerful aesthetic response, however, is the assurance of ultimate safety. The ascent on the rollercoaster can only be enjoyable and exciting when we are reassured from the outset that the downward plunge will reach a controlled conclusion and that we shall ultimately be delivered safely back to our starting point. Burke continues:

Whatever is fitted in any sort to excite the ideas of pain, and danger, that is to say, whatever is in any sort terrible, or is conversant about terrible objects, or operates in a manner analogous to terror, is a source of the *sublime*; that is, it is productive of the strongest emotion which the mind is capable of feeling. . . . When danger or pain press too nearly, they are incapable of any delight, and are simply terrible; but at certain distances, and with certain modifications, they may be, and they are delightful, as we every day experience.[29]

Burke and his contemporaries agreed that the sublime experience raises the observer to a higher rank and that such aesthetic responses can stimulate ethical insights pointing the way to ideal being. As the Scottish philosopher Hugh Blair noted: "Wherever, in some critical and high situation, we behold a man uncommonly intrepid, and resting upon himself; superior to passion and to fear; animated by some great principle to the contempt of popular opinion, of selfish interest, of dangers, or of death; there we are struck with a sense of the sublime."[30]

Yet the moral and intellectual reassurance engendered by the sublime can shift dangerously into an immodest overestimation of our human powers. The build-up to the Iraq war in 2003 can be viewed in precisely these terms. Indeed, the presentation of U.S. Secretary of State Colin Powell to the U.N. Security Council made recourse to a powerful sublime sentiment of absence in arguing that Saddam Hussein's "weapons of mass destruction" were poised to bring a terrible havoc to the world. In his account of the Al Musayyib chemical complex, Powell pointed to the emptiness of "bulldozed and freshly graded earth" as convincing proof of apocalyptic intent, in language redolent of Burkean horror: "The Iraqis literally removed the crust of the earth from large portions of this site in order to conceal chemical weapons evidence."[31] (See Fig. 3) The morbid spectacle of "shock and awe" was to follow.

In the context of the Holocaust, Philip Shaw has suggested that to the National Socialist ideologues, the figure of the Jew was the "surplus object in which, precisely, the Idea [of the unity of party and people] cannot be fully represented. . . . Simultaneously fascinating and repulsive, the hideous sublimity of 'the Jew' signifies the inability of fascism to be anything other than a fractured or empty totality."[32] Might it be that the "Axis of Evil" and the fear of weapons of mass destruction performed a similar task for the neoconservative agenda of Republican America in the mid-1990s?

Why pursue this thankless topic that defies close definition and is grounded generically on the anxiety of nothingness: a nothingness that can be physical, metaphorical, or spiritual? Kant offers an answer to this question when he writes, "Human reason has the particular destiny in one category of cognition: that it is burdened by questions that it cannot ignore; for by its nature reason itself imposes these questions on human reason, which human reason cannot answer, for they exceed all faculties of human reason."³³

Acknowledgments

I am most grateful to my son, Max Whyte, University of Chicago, for his critical reading of this text and to my colleague Mark Dorrian for pointing me toward the White House website depicting the Al Musayyib chemical complex.

Notes

1. Paul Crowther, *The Kantian Sublime: From Morality to Art* (Oxford: Clarendon Press, 1989), p. 2.

2. Edmund Burke, *A Philosophical Enquiry into the Origin of Our Ideas of the Sublime and Beautiful* (1757), J. T. Boulton ed., (London: Routledge and Kegan Paul, 1958), pp. 39–40.

3. Immanuel Kant, *The Critique of Judgement*, §23.2, trans. Donald W. Crawford, in Crawford, *Kant's Aesthetic Theory* (Madison: University of Wisconsin Press, 1974), p. 99.

4. Slavoj Žižek, *The Sublime Object of Ideology* (London: Verso, 1989/2008), p. 229.

5. Immanuel Kant, *The Critique of Judgement*, trans. James Creed Meredith (Oxford: Clarendon, 1969), p. 111.

6. Kant, *The Critique of Judgement* (1969), p. 190.

7. The appropriate Kantian quotation to accompany Friedrich's painting would be:

> The feeling of the sublime is, therefore, at once a feeling of displeasure, arising from the inadequacy of imagination in the aesthetic estimation of magnitude to attain to its estimation by reason, and a simultaneously awakened pleasure, arising from this very judgement of the inadequacy of the greatest faculty of sense being in accord with ideas of reason, so far as the effort to attain to these is for us a law (Kant, *The Critique of Judgement* [1969], p. 106)

8. Sigrid Hinz, *Caspar David Friedrich in Briefen und Bekenntnissen* (Berlin: Henschelverlag Kunst und Gesellschaft, 1968), p. 89.

9. Andrew Wilton, *Turner and the Sublime* (London: British Museum Publications, 1980), p. 74.

10. See Friedrich Theoder Vischer, *Über das Erhabene und Komische* (1837), reprinted as *Über das Erhabene und Komische und andere Texte zur Ästhtetik* (Frankfurt: Suhrkamp, 1967).

11. Crowther, *The Kantian Sublime*, pp. 164–165.

12. Nicholas Taylor, "The Awful Sublimity of the Victorian City," in H. J. Dyos and Michael Wolff, *The Victorian City: Images and Realities*, 2 vols. (London: Routledge and Kegan Paul, 1973), p. 434.

13. David Hume, *A Treatise of Human Nature* (1739/1740), quoted in Walter John Hipple, *The Beautiful, the Sublime, and the Picturesque in Eighteenth-Century British Aesthetic Theory* (Carbondale: Southern Illinois University Press, 1957), p. 43.

14. Vischer, *Über das Erhabene und Komische*, p. 155.

15. Gary Shapiro, "From the Sublime to the Poetical," *New Literary History*, 16 (1985), p. 216.

16. Friedrich Hölderlin, *Poems and Fragments*, bilingual edition (London: Routledge and Kegan Paul, 1966), pp. 372–377.

17. Martin Heidegger, quoted in Karl Heinz Bohrer, "Am Ende des Erhabenen," *Merkur*, 43 (September/October 1989), p. 746.

18. Kant, *The Critique of Judgement* (1969), p. 100.

19. Barnett Newman, "The Sublime Is Now," *Tiger's Eye* (October 1948), p. 53.

20. Jean-François Lyotard, *The Postmodern Condition: A Report on Knowledge*, trans. Geoff Bennington and Brian Massumi (Manchester: Manchester University Press, 1984), p. 79.

21. Paul Klee, "Gedichte," quoted in Wolgang Welsch, *Ästhetisches Denken* (Stuttgart: Reklam, 1990), p. 89, n. 24.

22. See Wassily Kandinsky, *Concerning the Spiritual in Art* (1912), trans. Michael T. H. Sadler (London: Tate Publishing, 2006), pp. 25–26:

> And when we rise higher in the triangle we find that the uneasiness increases, as a city built on the most correct architectural plan may be shaken suddenly by the uncontrollable force of nature. Humanity is living in such a spiritual city, subject to

these sudden disturbances for which neither architects nor mathematicians have made allowance. In one place lies a great wall crumbled to pieces like a card house, in another are the ruins of a huge tower which once stretched to heaven, built on many presumably immortal spiritual pillars. The abandoned churchyard quakes and forgotten graves open and from them rise forgotten ghosts. Spots appear on the sun and the sun grows dark, and what theory can fight with darkness?

23. Kandinsky, *Concerning the Spiritual in Art*, p. 77.

24. Karsten Harries, *The Meaning of Modern Art* (Evanston, IL: Northwestern University Press, 1968), p. 45.

25. John Milbank, "Sublimity: The Modern Transcendent," in Regina Schwartz (ed.), *Transcendence: Philosophy, Literature, and Theology Approach the Beyond* (London: Routledge, 2004), p. 212; also quoted in Philip Shaw, *The Sublime* (Abingdon: Routledge, 2006), p. 4.

26. Immanuel Kant, *Geschichte und Naturbeschreibung der merkwürdigsten Vorfälle des Erdbebens, welches am Ende des 1755sten Jahres einen grossen Teil der Erde erschüttert hat* (1756).

27. See Gene Ray, *Terror and the Sublime in Art and Critical Theory* (New York: Palgrave Macmillan, 2005), pp. 51–59.

28. Burke, *A Philosophical Enquiry*, p. 38.

29. Burke, *A Philosophical Enquiry*, pp. 39–40.

30. Hugh Blair, "Lectures on Rhetoric and Belles Lettres" (1783), in Andrew Ashfield and Peter de Bolla (eds.), *The Sublime: A Reader in British Eighteenth-Century Aesthetic Theory* (Cambridge: Cambridge University Press, 1996), p. 215.

31. See "Iraq: Denial and Deception: U.S. Secretary of State Colin Powell Addresses the U.N. Security Council," White House website (www.whitehouse.gov/news/releases/2003/02/20030205-1.html#). This transcript was posted directly after the speech by Colin Powell to a plenary session of the United Nations Security Council on February 5, 2003. It is no longer available on the White House website but can be found at http://georgewbush-whitehouse.archives.gov/news/releases/2003/02/20030205-1.html#26

32. Shaw, *The Sublime*, pp. 139–140.

33. Immanuel Kant, "Erklärung in Beziehung auf Fichtes Wissenschaftslehre," *Allgemeine Literatur Zeitung* (Jena), Intelligenzblatt no. 109 (August 28, 1799), pp. 876–878.

2

Affective Foundations of Creativity, Language, Music, and Mental Life

In Search of the Biology of the Soul

JAAK PANKSEPP

Our desire to know the world drives science. The search for beauty energizes the arts. Both are fertilized by the ancient affective energies that motivate human creativity. The pain and sadness of life, joy also, drive poets, musicians, all artists, to become much more than the physical organisms that they are. But how emotional feelings, from primal anger to tender love, are created within the human brain remains one of the most important and least addressed scientific questions in the mind sciences. Humanists have no better *knowledge* than do philosophers or neuroscientists of the nature of those volcanic emotional powers that drive their artistry, whether ordinary or sublime. But at least they are not as much in denial as many brain scientists are of their importance in mental economy. Our many affective feelings, basic tools for living, remain one of the great mysteries of existence.

Some feel that affect, the positive and negative feelings side of emotions, is just another kind of "cognitive" activity. They are wrong, at least at the primary-process level. Although emotion and cognition are completely blended in our higher psychological experiences, brain research and great art offer unique ways of distinguishing thoughts from the primary-process affects. They emanate from different regions of the brain. Our most intense affects emerge largely from ancient "basements" of the brain, and our most sophisticated cognitions from the cortical roof. They also operate by different principles, such as "information processing" versus "state-control regulation."[1] They share some brain chemistries, but others (e.g., certain neuropeptides) are uniquely emotional. Of course, cognitive and affective zones of the brain interact just as do all organs of the body. Also, there are some old cortical regions that naturally link emotional states to cognitive demands, but most neocortical regions are intrinsically designed not to process emotional feelings but to deal with the endless flow of information about the external world entering the mind through our sensory portals—the gateways for appreciating, and even finding awe in, the surrounding world.

All artistic creations require great cognitive skills, but works that do not stimulate our feelings communicate little. In art as in life, affects motivate cognitive richness like torches illuminate the darkness. Consciousness is colored and integrated by the ancient emotional systems of our brains.

The Affective Foundations of Mental Life

Without affect, we would not feel alive. Without affect, there is neither pain nor fun. Without positive affect, there are few reasons for living, and people in depressive despair often choose death over life. Affect is the source of all intimacy—the profound interpersonal giving of oneself exhibited by people who are in love with life and others. Affect encourages people to dig deeply into their biological "souls"—to find empathy, to sincerely communicate their major concerns, and to hope their depth of feeling is reciprocated. Affect is the fuel of the cognitive mind, allowing it access to the sublime. Affect is the muse that has a mind of its own.

Affects fill the mind with a cornucopia of desirable and undesirable experiential states that are hard to define objectively and hard to talk about clearly. Thus, neuroscientists shun affects in their science, and often in their personal lives. In part this is because raw affects are ancient idea-less forms of consciousness composed of brain and bodily processes of kaleidoscopic complexity—as dynamically beautiful as the universe envisioned through Hubble portals (see chapter 4). Regrettably, despite great progress, we do not yet have comparable mind-scopes for the brain. Most brain functions remain like "dark energy" to current functional brain-imaging technologies.

Affective feelings are the basic value systems of the brain/body. There are many distinct types of feelings. Some accompany major bodily disturbances (e.g., pain and fatigue); some reflect sensory pleasures and displeasures (e.g., from tasty delights to disgust); others gauge bodily need states (e.g., hunger and thirst). Perhaps most mysteriously, certain intrinsic brain-body states are intensely imbued with very special feelings of goodness or badness—the *emotional* affects, evolutionarily designed to allow organisms to respond distinctly to life-challenging world events.

How does neural activity create *emotional* affects? Neuroscience has been wary of such questions, but it it is now clear that the same areas of the brain that generate instinctual emotional behaviors are the ones that generate intense emotional feelings. This suggests a *dual-aspect monism* view of mental life, where complex neural systems that regulate visible emotional behaviors are largely the same as those that generate invisible emotional feelings. In the future, this may be the scientific gateway for understanding how feelings are constructed from fluctuating brain dynamics. In a metaphoric sense, this is not much different than physicists having to accept that light has both particulate and wave-characteristics.

Regrettably, many brain scientists (with increasing numbers of notable exceptions, e.g., Antonio Damasio, Oliver Sacks, Mark Solms) cannot find space for experience, especially something as fuzzy as emotional feelings, in the complex biophysical puzzle of the behaving brain. Some, perhaps most, are convinced that once we understand every neuron, every neurochemical, and every electrical spark in the brain, there will be nothing important left for subjective experience to do in a brain that *is* governed by physical laws. As the late Heinz Pagels said, with some dismay: "Most natural scientists hold a view that maintains that the entire vast universe, from its beginning in time to its ultimate end, from its smallest quantum particles to the largest galaxies, is subject to rule—the natural laws—comprehensible by a human mind. Everything in the universe orders itself in accord with such rules and nothing else."[2] Where, within this "vast universe," is there a place for mental feelings that is compatible with the "natural laws"? Many skilled neuroscientists and talented philosophers say nowhere—if subjective experience is anything, it is just the puzzling (epiphenomenal) end result of natural complexities.

Others, like myself, who subscribe to a dual-aspect monism perspective, suggest the evolution of affective feelings was nature's way of inserting intrinsic, albeit initially simple-minded values into the ultracomplex biological complexities of the brain. Despite heated philosophical debate, most neuroscientists still subscribe to a "ruthless reductionism" that insists we are *just* ultracomplex biochemical machines and nothing else. For most humanists, this does not seem right. Neuroscience may justly claim that our minds are completely biophysical, chemical (viz., physically material), but somehow, subjective experiences—the prime indicators of mental existence—arose from brain neurophysiological substrates. Perhaps that was achieved (although I deem it unlikely) through some kind of yet undiscovered "mind dust" that permeates the quantum universe. Even though most psychologists now agree with the once Freudian heresy that unconscious neural processes permeate the mind, most also agree that the brain has enough complexity to generate mind. There is room for consciousness within the brain, and it goes far back in BrainMind evolution.

Perhaps the first experiences that existed on the face of the earth were strong feelings. Ancient forms of pain and pleasure were surely among the first. Desire and despair were perhaps not far behind. As brains evolved, these rudiments of mind flowered into the infinite varieties of mental life that drive humanistic and artistic endeavors. Emotional feelings are the wellsprings of all creative urges. With the emergence of affective states, and their many uses, evolution gave us mental life that could eventually be self-aware, could appreciate the world, could see beauty in aspects of the world that could support survival—in the seeming endless varieties of flora and fauna (some we could eat, and some would eat us), sublime geological formations to take our breath away, weather patterns that could intimidate us, and other fluxes of the universe. With the advent of mind, life became much more than "crossed staves in a field, behaving as the wind behaves."[3]

Eventually, our capacity to feel ourselves and the world, ancient tools for living, opened up the possibility of sophisticated forms of learning, leading to ideas, thoughts, creativity, and eventually the desire to be closer to the angels than to the dust of the earth. The other creatures share many of our basic emotional feelings, but for one simple reason they are not as smart as we. They have comparatively modest neocortical "thinking caps." As a result, they do not play instruments, they do not paint, they do not write. Perhaps they have no emotions sublime. But they do experience their basic emotions intensely—playful joy, fearfulness, anger, lust, and many other desires.

Our massive expansion of the neocortex allowed us to be pioneers into cognitive and artistic realms, but its sudden mushrooming was no great biological achievement. Once we had discovered cooking, that allowed our cranial plates to expand (i.e., no longer constricted by massive masseter muscles needed to grind our food), cortical expansion was based on a multiplication of "computer chip"-like repetitive structures of three thousand neurons or so-called *cortical columns*. Once the complex internal structure of columns had evolved, it required only a few genetic tricks to lead to proliferation, almost like a hopefully benign tumor (one that could create great good and great evil), based on the principle "more of the same, please." This endowed us with a better capacity for symbolic thinking and communication than any other animal and a greater capacity to manipulate the world. It made us smart.

Our human neocortical expansion occurred rapidly, in an evolutionary twitch of time, reaching a point now where further growth would impair infants' ability to enter the world through the narrow human birth canal. The ancient subcortical waking, dreaming, emotional, motivational, and memory systems, all sub-cortically concentrated, required much more time and evolutionary "effort" to construct than did the expansive computational space that allowed the human cognitive mind to flourish. Contrary to all-too-common evolutionary speculations, most of the functions of those higher regions, even visual competence, are acquired through use and practice (namely "epigenetics" the ability of environment to guide what genes do) rather than rigidly genetically specified in the nucleotides of DNA.[4]

Modern cognitive neuroscience, with a fascination with the "computational theory of mind" has been focusing mostly on the functions of the higher regions of the brain, along with the sensory portals that connect human minds to the world and the motor abilities that allow us to work for a living. Because of many new human brain imaging technologies, there is also now much interest in how the brain processes emotion-related information (e.g., angry faces) but comparatively little work on emotional feelings.

The Passions of the Mind: The Ancestral Voices of the Genes

So what are our basic emotional systems—systems that generate feelings beyond our bodily needs, beyond our bodily itches, thirsts, and hungers? There, at the transition of the bodily feelings and the completely internal feelings of the brain, we have hormonally driven brain systems for lust, somewhat different in men and women, sometimes (perhaps always) mixed up to some extent such that male-typical affective mentalities can exist in female bodies and female-typical mentalities in male bodies. (See Fig. 4) Many are hermaphrodites in their minds, some in their bodies. But beyond the biology, there are the vast complexities of individual learning and culture, realms of mind that are always supported from below by the impish orchestra of primitive biological feelings we still share with other animals.

Beside the primitive urges of anger and fear, there are brain systems for maternal devotion that so readily generate caressing care and loving hugs, without which the milk of human kindness and empathy do not flow. (See Fig. 5) And for those who love evolutionary insights, it is delightful that this basic emotional system of mammalian brains relies on molecules that allow milk to be manufactured in female breasts (prolactin) and molecules that squeeze milk from the breast (oxytocin) as infants cry and grapple to suckle on the fountain of life. These molecules, which help mediate social comfort and confidence can also melt sadness, just like opiates, but they are not dangerously addictive. In fact, oxytocin activity in the brain makes opiates and presumably endogenous opioids, which mediate mother-infant attachments, less addictive by not allowing the brain to become insensitive (develop tolerance) to them.

All mammals, so dependent on maternal devotion to survive, have extensive neural networks to mediate the panicked-lonely feelings of social loss. These systems for sadness, (See Fig. 6) first mapped out in animals using localized brain stimulation, are powerfully activated by executive stress chemistries of the brain, such as corticotrophin-releasing factor, and they are soothed by our brain opioids and oxytocin. These ancient emotional systems are very similar in all higher animals, whether guinea pigs or humans, and even birds. And even invertebrates, like crayfish, like addictive opiates and psychostimulants, but we do not know that they form social bonds with each other. In all the social vertebrates, feelings of psychic pain that arise from social loss are controlled by similar neural "highways" and the same chemistries. We also now know that human sadness is accompanied by diminished activities in human brain opioid activity.[5]

And just as specific neural systems exist for grief and sadness, there are others for play and joy. The play urges of the brain are governed by ancient mammalian circuits that foster positive social engagements and encourage young mammals to explore social possibilities and to learn the complexities and opportunities within their social worlds, all those nuances that cannot be coded in the genes. Human children also have strong desires for physical play, to see how much they can do to and with each other. And there are consequences for brain maturation: play is superbly effective in activating the genetic orchestra throughout the higher neocortical reaches of the brain that are not needed for animals to play. Children who fail to get enough of these joyous activities may experience all kinds of developmental problems. They may be more susceptible to depression later in life since their social-joy systems will not have grown vigorous with use. Play- and attention-starved children may be labeled with attention deficit hyperactivity disorders and given powerful addictive drugs (psychostimulants, e.g., amphetamine) that reduce their playful urges; all the medicines used to treat such impulsive kids dramatically reduce playfulness. Societies that do not allow abundant free play for young children may eventually have more and more citizens whose social brains are impoverished—less capable of interacting gracefully and appropriately.

Interlude: Cross-Species Affective Neuroscience and the Search for New Mind Medicines

Because of the progress made in deciphering the basic neural circuits that control basic emotions in other animals, some of us believe that to understand the deep mysteries of human mental life, we must first understand their rudiments in other, kindred animals who share the evolutionary roots of our primal emotional nature.[6] It is now clear that we mammals have all inherited complex brain networks that control our fears, our angers, our sadness, and our social bondage—our love and hate for each other. As noted, the brain molecules that resemble the elixir of the poppy—endogenous opioids that the brain itself produces—can reduce all painful feelings, especially the sting of loneliness. Opiate addicts are driven by their painful loneliness to consume molecules that can engender feelings of loving satisfaction.

Many get hooked because they never found solace through the grace of supportive companionship. Physicians could use these molecules as antidepressants, especially the relatively nonaddictive ones such as buprenorphine, but government agencies and society remain terrified of such powerful knowledge and medicinal possibilities, and perhaps what they may tell us about our deeper nature—that ultimately love is addictive.

Would it diminish our sense of human values and responsibilities to develop mind medicines like this as we come to understand that our feelings ultimately reflect vastly complex neurochemical tides within the synaptic jungles of our souls? Scientific creativity has now manufactured new pain-soothing molecules, like buprenorphine, that are less addictive than the narcotics Mother Nature constructed. These are invaluable tools for the humane treatment of opiate addicts. As we come to understand our emotional nature, many others neuropeptides will be found to modify each and every one of our basic emotions. This will open up a new era in the development of mind-medicines and mind-enhancers, perhaps new molecules that can help transport the mind into awful and awe-filled spaces filled with feelings heavenly, horrifying, and sublime.

Seeking: The Most Primal of Emotional Systems

Underneath all the desires of human and animal hearts, there is a system that energizes the primal urge to engage with the world: a "seeking system" that is fired by dopamine, a molecule that is essential for all aspirations and addictions. The exhilarating feelings promoted by brain dopamine produce an ecstatic aliveness, an urge to engage with all aspects of the world. It is one of the major fonts of creativity. It is critically important for libido—the "life force" that sets the stage for all aspects of human aspirations and cravings. When the mind is energized in this way, it is eager to do all kinds of things. People often feel they can accomplish anything. Imagination can become obsessively wide as dopamine surges in the brain or become narrow and delusional when the mind is chronically overstimulated.

Dopamine-energized urges to seek can help all mammals to solve their life problems and at times to be overly creative, generating delusional causal connections between correlated events. The neural currents of creative energies, whether of mice or of men, flow best within the higher reaches of the mental apparatus when present moments are driven rapidly by dopamine-engendered feelings of euphoria and remembrances of things past. When dopamine cells stop firing, our sense of time slows down. Dopamine is essential for joy in living, as was discovered when animals were found to eagerly increase arousal of this system, whether with drugs or with localized electrical stimulation of the brain. Such urges and cravings can now be monitored through animal play and laughter.[7] When this system is depleted, animals and humans become deeply depressed. When doctors stimulate this system artificially, depression can be reversed.

The seeking system is necessary for all other emotional systems to function optimally. When brain dopamine activity becomes imbalanced, as already noted, it is also one of the major sources of drug addictions, depression, and illusions. When in overdrive, it is one of the sources of psychotic delusions as well as the seeking of spiritual heights. Without dopamine, one falls into an endless waking-sleep, a Parkinsonian stillness, from which awakening was not possible until neuroscientists deciphered that chemistry of the brain. This emotional system is essential for all other emotional systems to function . . . and cognitions too, perhaps even our passion for language.

Affect as the Source of Music and Language

Without affect, we humans would have little to talk about, and no special reason to reach out to others. Affects motivate our urge to play and to speak: when one of the highest brain regions that encodes sadness, grief, and social bonding (the anterior cingulate) is damaged, people descend into *akinetic mutism*. Such unfortunates retain the physical capacity to speak, but they have no urge or wish to communicate. All our cognitive activities are energized by our passions. Our language is motivated by our social emotions. Might it even be possible that our capacity for language was first driven, in evolution, by our social-emotional nature—basic communicative urges that were codified in our melodic intonations long before the emergence of language? Our own basic emotional sounds, from the moan of agony to shrieking for joy, passed down to us from our basic animal nature, may have served as preadaptations for ancestral singing and thereby the cultural inventions of music. Indeed, that proto-musical nature may have been the transitional passage to the evolutionary emergence of language. This idea, if true, may also help us understand the strong linkage between the cadences of poetry and music. Thus, rather than there being a distinct "language instinct" where music is deemed nothing more than "cultural cheesecake," perhaps our musical-poetic nature was a prerequisite for the invention of language.[8] Perhaps our musical nature is constrained much more by our genes than our capacity to acquire propositional language? Let us pause to consider this strange idea.

Clearly, human infants exhibit musical interests and abilities before they exhibit speech. Perhaps their capacity to process auditory information with sing-song musical tempos and inflections anticipates the acquisition of speech. Let us look at this strange possibility in component parts:

- Animals communicate with sounds, but only affectively and with greater subtlety than usually imagined. Even young rats make a laughter-type sound when tickled: if you tickle them well, they will bond with you and delight in your companionship.
- Some kind of "proto-musical" ability precedes language in human development, and mothers use a special, higher pitched rhythmic intonation, called "motherese," to encourage engagement and communication in their infants.
- Music is the most sophisticated human "language" of emotions. Its emotional power is dependent on the arousal of subcortically situated emotional systems.
- Music and language capacities overlap, being tightly coupled in the brain.

I find feelings sublime in the fascinating possibility that the preadaptation of primal emotional communication drove the cultural discovery of language. Perhaps there is no "language instinct" but only an "affective communication instinct" that led humans to discover language, because of the evolutionary conjunction of strong social passions, from grief to playfulness, existing in brains that finally had enough random-access memory space, enough multi-modal neocortex to associate information from the various senses, to have symbolic ideas. Feelings still permeate language and thrive in most lovely forms, such as poetry. And social emotions permeate all artistic endeavors.

But how does one sift fact from fancy when one is seeking answers to evolutionary issues lost in the mist of time? By looking for evidence in living genes and brains. Though robust genetic evidence remains scarce, clearly our appreciation of the emotional power of music resides in more ancient regions of our brains than do our abilities to appreciate the vast complexities of ideas that arise from the human neocortex.

Affects and Artistic Passions: Musically Induced Skin Orgasms

Some of the peak artistic experiences of the human mind have finally been captured in the butterfly net of neuroscientific inquiry. Since music is one of the most emotionally powerful artistic forms, it provides a fine opportunity to seek such knowledge. Music, perhaps better than any other art form, reflects our deep social-emotional nature. Of the vast spectrum of musical experiences spanning the full gamut of human emotions, love and loss, and joy and sadness are ever-prominent themes.

Perhaps the most dramatic and consistent bodily effects induced by music are the feelings of "shivers" or "chills" many people experience when intensely moved by the emotional power of music. Such "skin orgasms" are commonly evoked by bittersweet songs of unrequited love, loss, and longing, reflecting very personal social emotions. But also chills can be aroused by patriotic pride stirred during the commemoration of lost warriors, highlighting our broader social nature. High-pitched sustained crescendos, arising from the background of a grief-filled score, "pierce the soul," so to speak, and seem ideally suited for evoking chills.

Such autonomic arousals, measurable peripherally as galvanic skin responses, remind us of our social nature. They reflect how definitively ancient subcortical brain regions are the wellsprings of our social nature. Brain imaging studies have also shown abundant arousals in ancient socioemotional regions of the brain when people listen to chill-evoking music.[9] This raises the possibility that we might be able to create some kind of "proto-music" for other mammals and birds by utilizing and stylizing their own species-typical emotional vocalizations. This would be a sublime contribution for cross-species relations, improving our mutual welfare.

Why, from an evolutionary perspective, would we have such moving and chilling responses to music, feelings that transport us toward ecstasy? The answer may lie in the fact that the acoustic properties especially effective in triggering "skin orgasms" are sounds that resemble the separation cries of babies—the primal care-soliciting signals that promote social concern and attentive empathy, especially in mothers. Musically evoked chills may arise from the ability of certain emotional, musically rendered sounds, resembling separation cries, to access the evolutionary roots of social pain. These sounds reflect and communicate the painful emotional impact of social loss. Such feeling may partly arise from the ancient neural "soil" that controls how warm or cold we feel. Perhaps the sound of a lost child sends shivers across our skin because to be alone is to feel cold. This may be nature's way to motivate animals to restore social "warmth."[10] When that fails, the gateway to depression is opened wide.

Music that evokes chills often blends a wistful sense of loss with the possibility of reunion and redemption. Such aesthetic experiences hark to the foundations of our humanness—our profound loving attachments and dependencies, our massive social interrelatedness to other people, feelings that are partly created by endogenous opioids and oxytocin within our brains. All this occurs in ancient emotion-generating regions of the mind, perhaps where our "souls" are concentrated, among ancient neural networks that control our deepest human feelings. These circuits existed long before the evolution and massive expansion of our cortical thinking caps that eventually had enough cleverness to invent musical instruments to generate pure sounds not found in nature (tones), and the wherewithal to develop musical notations and other cultural conventions to artistically amplify and archive auditory mysteries. Perhaps one of the most puzzling phenomena is our ability to appreciate the self-similarity of sound across octaves, a perceptual phenomenon that still has no compelling evolutionary explanation and seems not to exist in other animals.

It is wondrous to behold how the complexities of culture have been built upon the foundation of our solid animal nature. Without our animalian emotional circuits, we could not be the passionate, artistically sophisticated creatures that we are . . . creatures with lots of "soul."

The Neurobiology of the Soul

So what does it mean to say, "This music touched my soul"? Perhaps we mean the chilling tremors that run across our body and mind when artists hit those shivery sweet spots of our emotional circuits that now allow us to experience ancestral feelings as sublime delight. I believe that among those ancient reaches of the brain we will eventually find the human soul. And if we do, it should not be all that different from the souls of kindred animals with whom we remain fortunate enough to share the earth. Only recently have neuroscientists had courage enough to entertain the possibility that the human brain contains neural functions that may deserve names such as the "core self" or "the neural soul." We may need to consider such perennial mysteries, barely captured by words, to envision how raw emotional feelings emerge from the complexities of neural activities.

Our current understanding concerning the fundamental nature of the "self," at psychological and physiological levels, remains modest, but scientifically testable ideas are being advanced. One is that a "core self" exists in ancient regions of the brain, such as medial regions of the brainstem, concentrated in brain areas such as the periaqueductal gray, where all basic emotional systems converge.[11] What is transpiring in such brain regions is not just "information processing" but the regulation of brain and bodily "states." These "state functions" exert global pressures on the mental apparatus, probably reflected in various "attractor landscapes" sweeping across the brain like weather patterns, generating tempestuous emotional feelings. Some claim that such experiences do not exist in other animals. Nonsense! That makes no evolutionary sense. We mammals, all, are inheritors of emotional storms that are experienced.

Do these affective dynamics manifest animal souls? Although our cortical surfaces contain an enormous map of our body surface (distinct motor and sensory homunculi), there are other, more diffuse and less understood body maps in lower regions of the brain. Mappings of our visceral organs, so vital to life, along with simplified body schemata generating instinctual-emotional actions, exist in ancient brain regions below the neocortex. Here is where our raw emotional feelings are forged from neural activities by yet unfathomed algorithms. These virtual body representations, deep in the brainstem, perhaps the first neurosymbolic body schema that emerged in neural evolution, may mediate those vital functions that, through centuries of cultural evolution and wishful hoping, perhaps paranoid delusions and fantasies, became symbolized by the concept of an immaterial soul. This comes down to us as Cartesian dualism.

Emotional feelings may arise from distinct dynamics induced by the various instinct-generating neural systems that converge on the ancient virtual body symbolized by neural circuits that map our viscera. Indeed, it is from these visceral brain regions that one can evoke emotional feelings by application of the lowest amounts of energy using localized neural stimulation. These are the brain areas where the smallest amount of damage inflicted to the brain eliminates consciousness, yielding persistent vegetative states. These are the brain regions where rapid eye movement dreams are generated in higher brain regions by neural circuits that are more ancient than the ones that allow us to be awake—a sublime paradox!

Neuroscientifically, this is as close to "the soul" as science has tried to reach. It leads to testable predictions: as a dying creature passes from life, these soulful brain regions, like the periaqueductal gray, should be among the last to die. Modern neuroscience has the tools to answer such momentous questions. Were this prediction to hold true, it would make this investigator have a sublime shudder. It would be as moving as when we discovered laughter in rats.

Human Souls, Animal Souls

Do other animals have souls? Do they have creativity? Art? My wife is fond of saying that all animals are spiritual creatures. They live in the dynamic present and not as much in past and future as we humans are prone to do. They may not reflect upon their feelings, and hence may not be "aware" of themselves in the sense that we are aware, but they are not unconscious zombies, as all too many neuroscientists, steeped in skepticism, still believe. The continuing belief that other animals do not feel affective "livingness" is a corrupt philosophical position that has failed to weigh the modern evidence, pro and con . . . a residue of Cartesian dualism. Their lives, as ours, are vibrantly full of affects that guide their life decisions.

When veterinary students keep one simple thing in mind as they learn to heal animals, half their learning is done: bring sick and injured animals back to their comfort zones. Perhaps medical students should remember that lesson, too. At some moment, perhaps indeterminate, the application of stressful modern technologies should cease, and human empathy and wisdom must prevail, to allow passages from this life with grace. We are biochemical creatures, but the greatest and most mysterious gift of nature is that we, and our fellow animals, have strong feelings about many aspects of living, many self-created. But our brains do have intrinsic values. All of us, humans and animals alike, exist in our personal bubbles of affective consciousness, with galaxies of feelings that glimmer through our mental apparatus and, through our emotional actions, to those of others as well. Some of our basic emotional circuits resonate with empathy, evoking maternal care when we are pierced by the sound of panicked grief.

Human psychiatry may eventually relearn this lesson and realize that between the diagnostic categories and our semi-effective, serendipitously discovered mind medicines, there are emotionally confused people, often angry, scared, sadly aching, or libidinally depleted. For new and more specific mind-medicines, there needs to be a robust science of the neurochemical landscapes of the affective mind. And the only way we can have that is by studying the anatomical, physiological, and neurochemical details of emotional systems in our fellow animals in new and sensitive ways that lead to clear affective predictions in humans. For instance, as already noted, tiny doses of opiates can melt human sadness and reduce the sting of grief by acting on separation-distress systems we share with the other animals. This knowledge has been available for decades, but it remains unused in psychiatry. And we have designed penal codes and ever more prisons for those who seek to alleviate distress by self-medicating with opioids. Legislators need to understand the brain, so they can weave a better and more understanding social fabric. Psychiatrists have to study our ancestral emotional minds in order to provide better care.

We can only seek such new knowledge through the wonderful psychic energies of our "seeking systems"—brain processes we share with all other vertebrates. This wonderful tool for living evolved from chemistries, like epinephrine, that originally stoked up our intracellular metabolic furnace (Krebs cycle) leading to a brain system that can illuminate and focus our minds in stressful situations (norepinephrine), and with one more biochemical step, this family of arousal systems helps us expect the future and develop plans (dopamine). To understand our higher minds, we must have a clear image of our ancestral mind. The brain is the only organ where one can see the evolutionary layering of feeling brains within our thinking brain.

So, once again, where is the *human* soul to be found? In the heavens? Unlikely. Up among the higher reaches of our brains? Perhaps. Modern neuroscience suggests it is wisest to seek the rudiments of the human soul among the lower reaches where our evolutionary relations to the other animals remain firmly evident. The soul is fundamentally affective. We know that the higher reaches of brains, by themselves, could never sustain consciousness, and certainly no emotional feelings, whether vernal, venal, or sublime. The lower animalian circuits have the power to keep us awake, alert, and full of passions on their own, initially with little help from the cortex, which at the beginning resembles a blank slate more than evolutionarily prepared modules for future cognitive activities. For instance, our speaking left cerebral hemisphere is not essential for language, as long as damage there occurs during the first few years of life. However, once we have matured and become dependent on the conceptual re-representations in the cortex, we lose most of our mind after cortical damage.

As life and culture program the higher regions of the mind, we rapidly reach a point where it can no longer function without them. From the outset, the lower parts of the brain are essential for our raw emotional lives, and subtle body representations permeate these ancient territories of mind, barely mapped by modern neuroscience.[12] Once we become familiar with these ancient aspects of our mental apparatus, we may develop scientific visions of what we are really talking about when we speak of souls. This does not mean that souls do not grow and mature within loving families and cultures—they do. There are many levels to our minds, to our selves . . . to our souls.

Within culturally untutored human brains, we may be able to more clearly see the ancient souls we still share with other animals, those purely spiritual creatures of the world who live not in worlds of persistent doubts and ambiguities. Those animal souls are not tortured by belief-based delusions, not captivated by associations that exist nowhere else but within the ideational higher cognitive reaches of human brains. But this may also be the reason our fine fellow creatures of the world are not pressured, by their ancient affective energies, to partake in creative artistries that have opened up gateways to feelings sublime in our own species.

What a work is man? We are creatures that can peer over the precipice that surrounds our little secure nest of knowledge and belief, into the abyss, and feel both terror and bliss. As we seek redemption, we recall those experiences in art!

Notes

1. Jaak Panksepp, "At the Interface of Affective, Behavioral and Cognitive Neurosciences: Decoding the Emotional Feelings of the Brain," *Brain and Cognition*, 52 (2003), pp. 4–14.

2. Heinz R. Pagels, *The Dreams of Reason: The Computer and the Rise of the Sciences of Complexity* (New York: Simon & Schuster, 1988), p. 11. For detailed summaries of basic emotional systems, see Jaak Panksepp, *Affective Neuroscience: The Foundations of Human and Animal Emotions* (New York: Oxford University Press, 1998), for Pagel quote see p. 344, n. 26.

3. T.S. Eliott, "The Hollow Men," 1925.

4. M. Sur, and C. A. Leamey "Development and Plasticity of Cortical Areas and Networks," *Nature Reviews Neuroscience*, 2 (2001) pp. 251–262.

5. J. K. Zubieta, T. A. Ketter, J. A. Bueller, JY. Xu, M. R. Kilbourn, E. A. Young, and R. A. Koeppe, R.A., "Regulation of Human Affective Responses by Anterior Cingulate and Limbic Mu-Opioid Neurotransmission," *Archives of General Psychiatry*, 60 (2003), pp. 1145–1153.

6. Jaak Panksepp, "Affective Consciousness: Core Emotional Feelings in Animals and Humans," *Consciousness & Cognition*, 14 (2005), pp. 30–80.

7. Jaak Panksepp, "Neuroevolutionary Sources of Laughter and Social Joy: Modeling Primal Human Laughter in Laboratory Rats," *Behavioral Brain Research*, 182 (2007), pp. 231–244.

8. Jaak Panksepp, "The Emotional Antecedents to the Evolution of Music and Language," *Musicae Scientiae* (special issue of *Musicae Scientiae*, 2009/2010, devoted to "Music and Evolution"), pp. 229–259.

9. A. J. Blood, and R. J. Zatorre, "Intensely Pleasurable Responses to Music Correlate with Activity in Brain Regions Implicated in Reward and Emotion," *Proceedings of the National Academy of Sciences of the United States of America*, 98 (2001), pp. 11818–11823.

10. Jaak Panksepp, "The Emotional Sources of 'Chills' Induced by Music," *Music Perception*, 13 (1995), pp. 171–207.

11. Jaak Panksepp, "The Periconscious Substrates of Consciousness: Affective States and the Evolutionary Origins of the Self," *Journal of Consciousness Studies*, 5 (1998), pp. 566–582.

12. See Panksepp, *ibid.*, and Jaak Panksepp, "Core Consciousness," in Tim Bayne, Alex Cleeremans, and P. Wilkens (Eds.), *The Oxford Companion to Consciousness* (Oxford: Oxford University Press, 2009), pp. 198–200.

3

Still Deeper

The Nonconscious Sublime; or, The Art and Science of Submergence

BARBARA MARIA STAFFORD

However late that may already be which we can grasp with aid of names that have been handed down, it is a piece of mastery—of giving shape to and bringing into view—something that went before and that is beyond our reach. . . . everything that man gained in the way of dominion over reality, through the experience of his history and finally through knowledge, could not remove the danger of sinking back—indeed, the longing to sink back . . .
HANS BLUMENBERG, Work on Myth

[One] way of phrasing this theory which I am arguing against is to bifurcate nature into two divisions, namely into the nature apprehended in awareness and the nature which is the cause of awareness. The nature which is in fact apprehended in awareness holds within it the greenness of the trees, the song of the birds, the warmth of the sun, the hardness of the chairs, and feel of the velvet. The nature which is the cause of awareness is the conjectured system of molecules and electrons which so affects the mind as to produce the awareness of apparent nature.
ALFRED NORTH WHITEHEAD, Concept of Nature

It seems striking that those two major interpreters of venerable Western humanistic and scientific traditions, Blumenberg and Whitehead, similarly invoke a sublime archaeology of unearthing. (See Fig. 7) Making the submerged emerge permits that which is excessive, or mentally beyond reach, to heave into view. This essay argues that what makes the early-modern version of the sublime still relevant across current disciplinary divides is that it instantaneously integrates what otherwise remains, in Whitehead's apt wording, "bifurcated." It does this by exhibiting how those "causes" become attached to, or enter into, our awareness. By the *sublime*, I refer specifically to that overwhelming psychophysiological intrusion which—like love's fury—transiently manages to merge the personal awareness of our affective and cognitive states with the otherwise concealed and impersonal neurophysiological mechanisms underlying them. With our resistance in shambles, we are also able to perceive how this subjective raw experience is actually conjoined with objective nature, that is, bound up with the tormented or roiled topography to which it secretly corresponds.

From Burke to Kant, the notion of the sublime entailed the sudden and involuntary stimulation of terror and pity, the spontaneous psychophysical reaction of viewers, listeners, and theater audiences to dark, uncultivated, and rugged landscapes, black crimes, and passionate scenes: climbing to the heights of love or sinking into the gulfs of remorse. In this material monism where mind and matter acted instantly as one, flesh crawled, hair rose, hearts pounded, and tears streamed, without the intervention of conscious reflection.[1] The overwrought universe in which the sublime operated was hyperdynamic: sensing, feeling, and knowing resulted from an eroticized gravitational process whereby physical body and material world were pulled together into a resistless conjunction.

Unlike the numinous Longinian sublime of Hellenistic antiquity, the formulation of the late-eighteenth-century material sublime is predicated on the art and science of the emergence of the submerged: the restoration of what has sunk from view. In the *Peri Hypsous*, Longinus contrasts what he considers to be the rather pedestrian, lowly type of invention that marshals facts into "the whole tissue of the composition" and "the well-timed flash of sublimity" that shatters everything like a bolt of lightning, revealing the full and elevated power of the speaker "at a single stroke."[2]

In contrast to this epiphanic apparition striking from above (and akin to the Enlightenment fascination for exposing and lifting the ruins of vanished monuments from out of the depths of the earth), the *modern* anguished sublime is more a material elevation from below. The eruption of novel or rare cultural treasures (as well as of a new type of cataclysmic insight) from under the lava finds a significant parallel in the material monism of much contemporary neurophilosophy and neuroscience. The neuroscientific collapse of the mind into the brain is mirrored in the mental and emotional confounding triggered by the sublime. The transfixed beholder—in the grip of nature's *terribilità*—is transformed into the very matter that is wholly occupying her senses.

The nervous system—when overstretched and overexpanded by extreme stimulation—becomes resistless to intrusion. Free access to the viewer's mind and feelings (by someone or something else) arises precisely when volitional attentiveness and emotional control are destroyed. Significantly, the disintegration or ruining of first-person consciousness by a powerful third-person "other" is both a biological and a psychological event: the shattering that perceptibly welds together the brain–mind continuum. This excessive situation can be usefully contrasted with Alfred Gell's concept of *agency*. To be sure, this notion of agency is also relational and context dependent but always in connection to a "me" who includes other persons, things, objects, even machines, only insofar "as I am a [controlling] agent with respect to it."[3]

In fact, I am arguing that such incursions *cause* the singular self to become distributed. It is while under the influence of these visual and sensory bursts that we directly *touch* the invisible mind–body, feeling its material oneness, just as the archaeologist palpably experiences the intense merger of past and present when holding an unexpected find. At that instant, whatever form our "identity" might have possessed a second before now becomes emptied out and filled by some larger or greater force obliterating any illusion of freedom of choice.

Importantly, this rhapsodic aesthetic, as well as neurological condition, embodies a master problem endemic to both the arts and the sciences. The sublime is nature's brutally material coup d'état leveled against the Winckelmannian rule of beauty conceived as rational, spiritual, harmonious, measured, balanced. It addresses that other, agonistic, side of nature and human psychology: the war of turbulent elements, the tragedy of annihilation, the celebration of emotional torment, the darkness of unknowing. Whether considering the obliteration of the self-conscious self—as in the dispersal of our unified awareness into a fractured landscape—or the dissecting of the holistic brain into its modular fragments, the key and still unresolved dilemma is the same.

The question becomes, what could possibly galvanize these heterogeneous components to form a unity, given the absence of any immaterial binding medium? Just as field archaeology puts disparate debris in *physical contact* with other conflicting debris, theories of brain localization are predicated on neural parataxis—the setting side by side of discrete dedicated brain regions from whose direct, if mysterious, contact self-consciousness is thought somehow to spring. Does our rich inner associative life emerge via synaptic junctions, neurochemical transmitters, synchronization of brain oscillations, creation of a global workspace, or Darwinian dynamics of reentry, or is it merely an epiphenomenon—a persistent illusion intrinsic to our biological machinery that must be ruthlessly reduced to its material parts?

In its modern incarnation, the upwardly directed vector of the ancient sublime became inverted to feature what was down below instead. For both its charismatic apostle, Burke, and its skeptical analyst, Kant, the sublime retrieval of traumatic experience was implicitly founded on the model of archaeological excavation. Bringing the distant, the remote, the alien to the surface was the perfect analogue for how nonconsciousness wells up into consciousness. Painfully digging out the remains of Herculaneum and Pompeii from the Vesuvian tufa, or scraping the vestiges of the Colossi of Memnon from the Egyptian sands, exposes cultural artifacts as soiled, that is, always already at one with the matter in which they inhere. And nature, in turn, is exposed as remote and radically material, stretching far beneath the ground. This earthly, performative nature is blindly autocratic, sublimely demolishing anthropocentrism. Humanity is knocked off its pedestal in a reversal of power relations. As Georg Simmel aptly remarks in his reflection on ruins, it is the wrath of nature that continuously overpowers the works of man, periodically "peeling off" the artifices of paint, and covering their metal, wood, or ivory ornamentation with the patina of age, to make those vainglorious artifacts its own again.[4]

Not surprisingly, then, the sublime finds its objective correlative in the subterranean drama of demolition. It is not accidental that the late eighteenth century was also an era enamored of nested cosmic epics: Dante's multistoried pit *Inferno*, Milton's fiery gulf of Hell, or Michelangelo's cavernous and scabrous *Last Judgment*. Grottoes, chasms, holes, cracks in the ground: all these barbaric and fissured "natural" ruins are the antithesis of Winckelmann's holistic, luminous, and Apollonian Hellenism.

They fly in the face as well of the placating picturesque mode—with its praise of pleasing decay, rolling vistas, and life lived on the sunny level.

The sublime, by contrast, stages nature as catastrophic, peopled by nonclassical Greeks and non-Augustan Romans. Think of Henry Fuseli's Iron Age heroes and Eddic rhapsodes, or William Blake's pre-Noachian Flood giants and chorus of Celtic Bards, or the vanished Germanic tribes buried in Caspar David Friedrich's neolithic tumuli. These mythic figures inhabit an anxiety-producing, extreme environment littered with monstrous jutting rocks, fathomless abysses, and cliffs that fall endlessly away. The viewer of such demonic drawings and violent paintings is plunged into a corresponding state of anxious expectation. The act of beholding is fraught with danger since we virtually mirror, and thus incorporate, that which is intrinsically destructive to our own psychic integrity. The sublime in art thus unnervingly highlights the phasic and episodic character of identity, that is, the fact that we apparently do not pay attention as we please. Rather, the conviction of our personal psychic intactness gets periodically wrecked by those self-ruining incursions coming from the outside.

We may well ask what made, and continues to make, this dramatic obliteration of the ego, or emptying out of the cave of the self, so attractive? Like the romantic longing to grasp prehistoric origins, primordial legends, or irrecuperable myths, the sublime offered a way to arrive at a radical, preepistemological condition *anterior* to all conscious acts of knowing.[5] As in Whitehead's "unbifurcated" philosophy of organism, a sublime thought is "a tremendous mode of excitement" dramatizing the unity underlying the diversity of the universe.[6] The theory of the sublime thus goes beyond any simple distinction between the natural and the cultural to foreshadow the mathematician/metaphysician's concept of the "superject." As this futuristic term suggests, Whitehead forecasts the emergence of a new entity under the sun, one capable of overriding the long-standing ontological dualism of subject/object.

The philosopher goes on to claim that when an "actual entity" encounters a "physical datum" and reacts to that thing which is not me, the intrinsically relational nature of all experience stands revealed. Bizarre neologisms notwithstanding, Whitehead is making an important point about the bridging of polarities. Not unlike what happens when one is under the spell of the sublime, for Whitehead, a particular individual comes into being only when he or she is put in intimate relationship with an external event that is radically unlike, and yet like, the experiencing subject. Further, he coins the word "prehension"[7] to indicate how the discrete self can come to contain other spatiotemporal "quanta," that is, how we are able to incorporate mentally and physically the many disjunctive singularities, operating at radically different scales, that we encounter daily.

Without mentioning the sublime, Whitehead nonetheless appears to have implicitly understood one of its major contributions to the contemporary intellectual scene. Like its key theorists, he adumbrates a theory of distributed consciousness. Selfhood is a dynamic relational process, and identity possesses a dominant directionality whereby the one becomes many and the many one. The *how* of the process is what makes it significant. To come together, the experiencing self must be drawn inward and *downward*, entering into and being penetrated by the otherwise unattainable situation. That is, the rising hierarchical order of Platonic ascent is inverted.

Recall that what drove the early-modern beholder of rugged scenery, eroded architecture, or melodramatic theatrical performances to bouts of ecstasy was the fact that such deliciously disturbing sights obliterated everything else inside or outside the mind. This possession resulted from being unwittingly and irresistibly forced to attend to something simultaneously outside and inside oneself without the labor of comparative and concurrent reflection. Stunned, the viewer abruptly becomes proximate with the viewed, escaping the constraints as well as the framework of Enlightenment ruminative rationalism.

Phrasing this aesthetic phenomenon in terms of the *science* of the sublime: our metabolism appears to drive ideation, and not the other way around. The sublime, defined as a profoundly disturbing and insatiable "enthusiasm," is a good example of "hot thought." The behind-the-scenes adjusting and regulating homeostatic system gets thrown out of kilter when overstimulated and strongly inflected by events that overwhelm its constant delicate balancing acts. One aspect of current neurobiological research, focused on proving that all thinking is infused by the emotions— running the gamut from cold to overheated and observed at the molecular, neural, or behavioral levels[8]—has much to learn from our fatal attraction to what is maddeningly unattainable and yet infinitely desirable. In addition, what makes the sublime of the humanist especially pertinent to that of the scientist is its construction as an undeniable *material* fact. Conversely, the sublime is a healthy reminder that more nuanced and subtle feelings do not just hit the spectator but require *practice*. They are, in fact, the result of social refinements and cultural reiterations.

The apparent naturalness, spontaneity, and convincing sensation of utter unavoidability—characteristic of the sublime— corresponds to the intensity with which we are seized by the sensible intuition that the mind is something internal, special, and different from the brain. In this scenario, our first-person consciousness resembles an eruptive separate faculty embedded in our depths—precisely because it so vehemently insists it is so.

This phenomenological mode of welding our private and buried flow of perceptions with the atomic continuum of nature represents the attempt to overcome what Isabelle Stengers has called the "proudly exhibited incoherence . . . of modern thought. Such a conflict pervades all domains where some kind of 'objective explanation'—be it neuronal, linguistic, cultural, political, or social or economic—may parade as a 'nearly sufficient' condition, arousing vigorous protests in the name of what would escape, or transcend, so-called objective explanations."[9]

As Antonio Damasio, Christoph Koch, Paul Thagaard, and Douglas Hofstadter (among many others) have argued, neuroanatomy, evolutionary biology, and cognitive psychology have variously shown how sensory and motor maps that represent the natural and artificial environments they interact with are enriched by ongoing evolution. Such augmentation or energizing by other entities and shifting situations furthers organismal complexity. Paradoxically, this emergent, newly participatory mind also tends toward solipsism since the neural networks recursively begin to speak to one another.

Hofstadter refers to these sets of interacting structures in the brain as constituting "the Gödelian swirl of self."[10] That is, what we take to be a deliberative act of consciousness is actually a fluid "dance of symbols" autogenerated deep inside our brain. Like the eighteenth-century proponents of the inevitable automaticity of the sublime, Hofstadter (following the neurophilosophy of Daniel Dennett) claims we are unable to see, feel, or sense the tiny invisible microforces that underlie either our selfhood or our thought. The "I-symbol" is just (sublimely) and profoundly locked in the feedback loop of human perception. Indeed, he speaks in the antiempiricist romantic fashion of Johann Gottlieb Fichte or Friedrich Wilhelm Schelling. Although our sense organs supply the brain directly, "I remain primarily myself" because "most of our perceptual input comes from our own perceptual hardware."[11]

Hofstadter's sublime "strange loop of selfhood" is also reminiscent of the visionary science deployed in Shelley's epic poem *Prometheus Unbound*—incorporating the dynamical ideas of Erasmus Darwin, Joseph Priestley, and Humphry Davy.[12] Like these romantic natural philosophers, we feel Hofstadter pulling farther and farther away from the surface of appearances, as if his starting point was the extreme, innermost limit of Shelley's celestial "all seeing Circle of the sun."[13] Once the sensory influx has gotten beyond the retina, perception has entered a sunless cave to become wholly an internal affair. In this metabolic underground, the "I" is merely a hallucination, an illusion, mirage, or episodic trick emerging from the autonomous nervous system.[14]

I have been arguing that the lowering of the sublime—already perceptible in embryo in the late-eighteenth-century archaeologically inspired model of unearthing—has become intensified in the downward plunge of much twenty-first-century scientific research. The neurosciences are not alone in favoring a steeply descending infinity. Perhaps this predominant focus on a single plummeting vector—encapsulated in the tunneling imaging probe—owes to the ever-receding vistas of genetics, to the spiraling perspectives of the nanoscale world, but, above all, to the compression of information technology. For it is information technology that made all this data mining and digging for information feasible. Information technology's progeny—electronic and digital imagery—is the result of shrinking, compacting, and condensing the cogent image until it vanishes into pixels. In principle, even though there are no directions in cyberspace, rampant fears of "cyber peril" torment our real lives.[15] How can one feel secure with profoundly embedded "logic bombs" and rogue "botnets" covertly threatening the World Wide Web? These buried intruders, capable of mounting sinister offensives against computational circuitry, are tellingly lodged in a virtual "space" whose sovereignty extends well below visibility.

Analogous to the cybernetic model of plummeting, research into the bottom-up roles of nonconscious perception continues to plunge deeper beneath the frontal and prefrontal lobes, probing under the cortical layers into the amygdala, thalamus, hippocampus, insula, and the nucleus accumbens. Left far behind, like waves rippling across the face of a pond, is research into the sorts of things that top-down consciousness actually does. I mean research that goes beyond recording whatever happens to be stimulating a particular sensation to ask how these multitudinous sensations are bound into the aesthetic unity of apperception.

Today, what seems to fascinate the neuroscientific community and the popular imagination as well are the roots of impulsive addictive behaviors,[16] the hypnotic images that spring up in dreams during both early and late (REM) sleep cycles,[17] and the subliminal seductions proffered by the electronic entertainment technologies of the blink. Not unlike the sublime, what all these unconscious behavioral guidance systems do is instantaneously link sight with its object, bypassing considered thought.

I believe this is the kind of experience Hofstadter refers to when he states that there are patterns imbued with fantastic "triggering" power. But instead of just extolling this automaticity, Thagaard, by contrast, argues that it is important to distinguish between when emotion-suffused thinking leads us to make good or bad, productive or nonproductive, decisions. In support of this distinction, he cites Damasio's analysis of somatic markers (i.e., those *covert* emotional signals tied in with previous emotional experience that get spontaneously reactivated in subsequent similar contexts). When somatic markers are passed on to the nucleus accumbens, the latter acts as a gateway, allowing only context-consistent behavior (determined by the hippocampal input to the nucleus accumbens) to pass through.[18]

To my mind, this fact highlights the critical need to understand the kinds of patterns that increase, not decrease, our cognitive options. By this I mean what sorts of images bring us up, out of the depths of our recursive and impulsive mechanisms, to summon us to their further consideration. The larger aim of my recent book *Echo Objects: The Cognitive Work of Images* was precisely to provide such a demonstration of willed seeing. I argued that brain scientists and neurophilosophers of every stripe cannot afford to ignore the cognitive *work* performed by nonillusory and hence nonautomatic genres: those that do not simply mesmerize but knowingly engage us.

The repertoire of images currently featured in scientific experiments or being mustered in support of scientific hypotheses about brain function must be seriously expanded to include nonillusory patterns of organization. If the sublime represents a second step beyond pure, appetitive responsiveness to engage in a fleeting and brusque relationship with the world, Whitehead's "concrescence"[19] would seem to be the logical third step.

The mathematician/philosopher defined this higher vision as one that would entail *knowing* how plural feelings and multiple forms actually crystallize into a cognizable or *facetted* unity. Like the exponents of the late-eighteenth-century sublime, he wanted to get beyond the Humean stringlike contiguity of associations to arrive at a composite integration.

But, paradoxically, Hume was on the right track. It is precisely in the *performance* of relationship that congruency-making intarsia unifies self and world, not by overwhelming or swallowing the one in the other. Whitehead, I believe, was searching for just such a revolutionary system of mosaicized planetary order. That sensory, hence aesthetic, order—based on taking the world apart to see what makes it tick and then piecing it together anew (i.e., not just giving the "impression" of reality)—is available, I propose, in the long tradition of anti-illusory "inlaid" artistic genres. These compressive, conspicuously seamed formats cut a wide swath in terms of the morphology of their raw stuff: historically, geographically, and in the variety of media they employ as building materials for combinatoric constructions.

To name just a few: recall the courses of blue tiles marching across the Babylonian Ishtar Gate; the floral or scenic wood intarsia lacing Gothic and Renaissance choir stalls, the grotesquely stitched-together emblem whose heyday did not end with Goya or even Manet; and on to Cubism's synthetic collage as well as to Dada's perplexing photomontage. The act of inlaying, which forces us to mentally inlay—thus following suit—obliges us to lift up out of our internal autopoietic processes and forge a relational selfhood instead. This creative interaction between an existing work and the conscious viewer, who continues to work upon it, produces true coknowledge. One might say with the Russian formalists, the viewer *specifies* the distinguishing features making such a connection formally possible.[20] Unlike the cognitive meltdown instigated by the thunderbolt of the sublime, the viewer of nonillusory patterns remains cognizant that she is actually fitting organism to world.

Notes

1. Jim Davis, "The Sublime of Tragedy in Low Life," *European Romantic Review*, 18 (April 2007), pp. 162–163.

2. Longinus, *On the Sublime*, trans. W. H. Fyfe, Loeb Classical Library 23 (Cambridge, MA: Harvard University Press, 1995), pp. 163–165.

3. Alfred Gell, *Art and Agency. An Anthropological Theory* (Oxford: Oxford University Press, 1998), p. 22.

4. Georg Simmel, "Die Ruine," in *Philosophische Kultur. Über das Abenteuer, die Geschlechter und die Krise der Moderne*, ed. Jurgen Habermas (Berlin: Wagenbach, 1998), pp. 118–124.

5. On myth, see George S. Williamson, *The Longing for Myth in Germany: Religion and Aesthetic Culture from Romanticism to Nietzsche* (Chicago: University of Chicago Press, 2004).

6. Alfred N. Whitehead, *Modes of Thought* (New York: Free Press, 1968), p. 36.

7. See Alfred N. Whitehead, *Adventures of Ideas* (New York: Free Press, 1967). See also Michael Halewood, "On Whitehead and Deleuze: The Process of Materiality," *Configurations*, 13 (2005), pp. 62–63.

8. On the wide range of emotional cognition, see Paul Thagaard, *Hot Thought. Mechanisms and the Applications of Emotional Cognition* (Cambridge, MA: MIT Press, 2006), p. 8.

9. Isabelle Stengers, "Whitehead's Account of the Sixth Day," *Configurations*, 13 (2005), p. 38.

10. Douglas Hofstadter, *I Am a Strange Loop* (New York: Basic Books, 2007), p. 234.

11. Hofstadter, *I Am a Strange Loop*, pp. 207–213.

12. See Melynda Nuss, "Prometheus in a Bind: Law, Narrative, and Movement in *Prometheus Unbound*," *European Romantic Review*, 18 (July 2007), p. 419.

13. Percy Bysshe Shelley, *Shelley's Poetry and Prose*, ed. Donald H. Reiman and Neil Fraistat (New York: Norton, 2002), stanzas 892–898.

14. For an extended critique of what I term an "extreme" type of phenomenology, see "Impossible Will?" chapter 6 in my *Echo Objects: The Cognitive Work of Images* (Chicago: University of Chicago Press, 2007).

15. John Markoff and Andrew E. Kramer, "U.S. and Russia Split on Cyber Peril," *International Herald Tribune* (June 27–28, 2009), pp. 1, 14.

16. See, for example, Benedict Carey, "Who's Minding the Mind?" *New York Times, Science Times* (July 31, 2007), D1, D6.

17. Rebecca Cathcart, "Winding through 'Big Dreams' Are the Threads of Our Lives," *New York Times, Science Times* (July 3, 2007), D1, D4.

18. Thagaard, *Hot Thought*, pp. 90–91.

19. Alfred North Whitehead, *Process and Reality. An Essay in Cosmology*, ed. David Ray Griffin and Donald W. Sherburne (New York: Free Press, 1978), p. 4.

20. Victor Erlich, *Russian Formalism. History-Doctrine*, 4th ed. (The Hague: Mouton, 1980), p. 171.

4

Pretty Sublime

ELIZABETH A. KESSLER

Since the Hubble Space Telescope's launch in 1990, astronomers have produced an array of compelling pictures from data gathered by the instrument. (See Figs. 8–10) The images show us a universe filled with glowing gases in vivid colors, galaxies swirled together in bands of light and dark, and innumerable stars. NASA and the Space Telescope Science Institute (STScI), the research center at Johns Hopkins University that manages the telescope, intend these images to reach an audience beyond the scientific community, and they have fostered programs, such as the Hubble Heritage Project, that release pictures designed to appeal to the public at large. The wide circulation and frequent reproduction of such Hubble images, which have appeared on everything from magazine covers to stamps, testify to their success.

On the surface, the enthusiastic embrace of Hubble images is a straightforward public relations triumph. NASA and STScI found a way to make images that matched popular taste while encouraging an appetite for more such pictures and continued support for multimillion dollar research projects. The Hubble's scenes of nebulae, galaxies, and star fields have reshaped what astronomers and the general public know about the cosmos and how both groups imagine it, although not necessarily in exactly the same fashion. While specialists examine the images alongside data and calculations, those of us without advanced degrees in astronomy depend almost entirely on aesthetic responses to the pictures when constructing a vision of the universe. Many astronomers share the public's appreciation for the images, but others dismissively refer to those made for public display as "pretty pictures," a phrase that diminishes their importance and raises questions about whether we should take the pictures seriously as aesthetic or scientific objects. Labeling them as "pretty" suggests that the images are lacking, although it is not immediately clear in what way. It is obvious, though, that prettiness does not have the lofty status of other aesthetic experiences, such as beauty and sublimity.

A more nuanced analysis of the Hubble Heritage images complicates how we judge them, contrasting the dismissive tone implied when scientists call them pretty with a much grander aesthetic claim. In their subject matter and appearance, the Hubble images employ the tropes of the sublime. Those who produce the images speak of a desire to evoke a response that shares much with the sublime experience. Should we then conclude that the Hubble images are sublime, or do they remain pretty? Considering prettiness as an aesthetic concept and its use by scientists alongside the Hubble Heritage images will help to answer this question. But, ultimately, neither term may fully account for our aesthetic response to the images because that response also depends on *how* they represent the universe, the nature of the relationship between the physical world and its portrayal in the Hubble images.

Prettiness carries a peculiar ambiguity that distinguishes it from other aesthetic responses. To judge something as pretty both acknowledges the sensual pleasures it provides and points to its inadequacies, a lesser position underlined when the word is modified as "more than pretty" or "merely pretty." The possibility of a higher aesthetic level that has not been reached lurks in every usage. A woman who is called pretty may feel genuinely complimented but also wonder why she does not measure up to the standards of beauty. The word's association with femininity emphasizes these hierarchical connotations. Even when pretty shifts from adjective to adverb, it tempers the meaning of the word it accompanies. A pretty good day could always be better. In all cases, prettiness makes no claim of distinctiveness. Although above the ordinary, prettiness is common and conventional enough to approach the banal. We might notice a pretty scene in passing, but it does not absorb our interest. It is something that catches our attention briefly and then disappears, remembered, if at all, as a vaguely pleasurable experience. Because the world is full of pretty things, they have little significance beyond momentarily enhancing our sensual enjoyment. Even if we reflect more deeply on something pretty, we do not expect to find anything underneath the decorative surface.

Prettiness might be considered a cousin to the picturesque, a term intended to accommodate experiences that were neither beautiful nor sublime. Yet the picturesque is not a lesser version of the sublime, as prettiness is of beauty, but a valued aesthetic achievement that combines elements of beauty and sublimity. By recognizing a correspondence between the natural world and idealized views fashioned by artists, as occurs in the picturesque, the viewer gains new appreciation for both nature and art. "Pretty as a picture," one of the few uncompromised uses of pretty, however, is synonymous with the picturesque, a favorable comparison between the world visually perceived and a more perfected representation.

Removing two tiny words shifts the tone quite radically: "pretty picture" does not imply the same admiration. What, then, to make of the phrase and its adoption by scientists? It is commonly used in conversation by scientists in all fields as well as in more casual forums—such as scientists' personal web pages—and it typically distinguishes between representations intended for display and those used for analysis. Sociologist of science Michael Lynch and art historian Samuel Y. Edgerton identified this division within the visual culture of science after conducting an ethnographic study of astronomers and digital image processing practices.[1] The authors listed aesthetics, promotion, color, and qualitative content as the markers of pretty pictures, and they associated scientific images with professional work, maps and graphs, and quantitative content. Often the most purely scientific images did not circulate beyond a small group of scientists who were directly involved with the research. Because Lynch and Edgerton were more interested in finding aesthetic qualities in scientific images—in the representations that the astronomers initially dismissed as lacking aesthetic attributes—they did not analyze the pretty pictures further or consider the qualities shared by all the images the astronomers produced. Increasingly, scholars have similarly focused on excavating the visual artifacts of science that never reach a larger audience.[2] Lynch and Edgerton made a valuable contribution by documenting the way scientists regard pictures and recognizing that they understood the production of pretty pictures as separate from their scientific pursuits. Bringing attention to forgotten images is an equally worthy project, but ignoring pretty pictures repeats the dismissive attitudes of the scientists.[3] The nature of their prettiness calls for further analysis.

Color is one of the defining characteristics of pretty pictures, especially in astronomy. The faintness of the distant nebulae and galaxies and the insensitivity of the human eye make it impossible to perceive any distinct colors in them. Even through the eyepiece of a powerful telescope, celestial objects look white or sometimes green. Photographic plates and digital arrays collect light over time, therefore increasing sensitivity, but astronomers usually observe with filters that limit the wavelength of light these detectors register, resulting in monochromatic images. To make color images, astronomers combine multiple exposures of the same object as observed through different filters. A composite made from red, green, and blue images will exhibit the full spectrum of visible colors. Such pictures then are the result of human intervention and choice. In many cases, astronomers do not need the different hues or a composite of several filtered images for their analysis. Aesthetics motivates the choice to make a color picture, and while conventions guide how astronomers assign colors to different exposures, the standards are not fixed, changing for different media, astronomical objects, and sets of filters. Not surprisingly, this allows for a remarkable range of color choices. For astronomers, the ideal color scheme conveys something about the physical properties of the object and increases its visual appeal.

Pretty pictures involve more than just color, though. At a fundamental level, scientists must shift how they think about what they are representing. The phenomena depicted become scenes that scientists compose instead of objects for analysis. Astronomers try to portray the celestial objects in a manner that they believe those outside the scientific community will find interesting and easy to understand. For example, images typically show an entire planetary nebula or galaxy rather than just a small section. The choice often contrasts with the focus of their research, in which scientists may study a tiny region and not even observe the entire form. Astronomers also make decisions about orientation, and, with the introduction of orbiting telescopes, they gained great flexibility in this regard. In the past, most astronomical images positioned north at the top and east to the left, a standard derived from the experience of looking up at the sky when lying on the ground with your head facing to the north. But cardinal directions are irrelevant for an orbiting telescope, and astronomers are now free to rotate images as they please. In the Hubble images, gaseous nebulae are often oriented so the columns point toward the top of the pictures or the layers of clouds pull the eye up and into the depths of the scene. Scientists also avoid arbitrarily shaped frames in their pretty pictures, making square or rectangular images. When a view of the Whirlpool Galaxy was published in the *Astronomical Journal*, it appeared as a jagged composite; the viewer can quickly determine that it resulted from combining several different observations. The version produced for wider circulation by the Hubble Heritage Project incorporated data from the Kitt Peak Observatory to fill the corners, creating a seamless image in a rectangular frame.[4] The latter choice aligns astronomical images with a wider tradition of picture making and the metaphor of a picture as a window to another world.[5] These compositional choices require astronomers to crop an image or combine several sources of data. To further enhance clarity, astronomers eliminate artifacts of the instrument or observing process that could draw the viewers' attention away from the key features of the representation. Together, these decisions result in an image that can look quite different from the raw version returned by the telescope.

While the practice of making pretty pictures for display has existed throughout the history of scientific representations, for several reasons scientists have dedicated more time and attention to them in recent decades.[6] The economy of big science requires constant bidding for more funding, often from government agencies that must in turn curry support from bureaucrats and the taxpaying public. With the intense competition for grants, any advantage becomes important. Eye-catching images may encourage those awarding the funds to look more closely at a proposal. Once approved and active, projects still require financial backing, and images, especially in the case of a high-profile and expensive endeavor like the Hubble Space Telescope, help to sustain interest in its continuation. The widespread use of digital technologies across all the sciences allows for broader access to the tools of image processing. The expertise of an engraver or photographer is no longer necessary for making images, although access to tools does not guarantee that every scientist can make a visually impressive representation.[7] The ease of displaying and circulating images online further encourages scientists' interest in pretty pictures. Journals charge authors extra fees to reproduce color images in printed editions of a publication, but electronic versions of the same journals display color figures without cost to the authors. In addition, scientists can easily and cheaply post their favorite pictures on personal websites.

Astronomers refer to all images made for display, whether printed in a scientific journal or promoted through a NASA press release, as pretty pictures, but the possible audiences for these images are extremely broad and varied, including other scientists, government officials, representatives from funding agencies, the media, and the general public. While all pretty pictures combine pedagogy with aesthetics, the intended audience influences which of these receives more emphasis and introduces fine distinctions. An image made for a scientific journal must clearly support the authors' claims. One for a larger audience cannot dispute the scientific content, but it places a higher priority on attractiveness, often resulting in a more polished representation. The audience can also change who produces the image. In some cases, as with the Hubble's well-known image of the Eagle Nebula, the same pretty picture will reach all of these groups, and the astronomers who proposed the observation and wrote up the research paper will craft the image. In other instances, such as that of the Whirlpool Galaxy, multiple versions circulate with subtle, but significant, differences. In addition to the team of astronomers who requested the observation, image specialists from NASA's public relations offices or special groups, such as the Hubble Heritage Project, will develop a new image—a prettier version because it focuses even more on visual appeal—from the same data set.

Over the history of the Hubble Space Telescope, NASA and STScI recognized the value of images that emphasized aesthetics over pedagogy. Despite this, the impetus to ensure the regular production of pretty pictures did not come, as one might have expected, from the press offices but from two staff astronomers at STScI, Keith Noll and Howard Bond, who took an interest in developing attractive images from Hubble data after seeing some of the first headline-grabbing examples. They wanted to ensure that images of similar quality were a regular output of the telescope because they believed such images would be the lasting legacy of the telescope: a means to repay the public for the investment already made, a way to encourage continued support for astronomy and science, and a tool for inspiring future astronomers and scientists. But, as detailed below, not all their motivations were entirely pragmatic. Because Noll and Bond had a sense of how much planning and effort such images entailed, they knew that not every observation would yield a colorful and well-composed picture and not every astronomer would have the skill or inclination to make an image from the data. But the pair felt confident that they could craft Hubble data into spectacular pictures, especially with supplemental time on the telescope and the help of competent specialists in image processing.[8]

The duo solicited the approval of STScI's director for what they called the Hubble Heritage Project and then assembled a team to join them in the endeavor.[9] Several people have worked on the project during its history, but two members, Lisa Frattare and Zolt Levay, have made critical contributions to the aesthetic sensibilities of the group. Frattare and Levay do the bulk of the image processing, and they have widely shared their methodology with others involved in producing images. Levay receives particular credit and praise from other members of the group, as well as staff at STScI and NASA, because he often makes the decisions about contrast and color, choices that distinguish Heritage images from other astronomical pictures. Interestingly, none of these key members of the group had significant experience or training in art or public relations before beginning the project. Their notion of how the images should look is informed by scientific standards and a general awareness of artistic principles.

Since the group released its first set of pictures in October 1995, they have compiled a gallery of striking images, adding an average of one each month. They used the large archive of Hubble data for about half of their images and collected new observations for the remaining ones. The group solicits observing time from an allotment known as the director's discretionary time, which is time left open in the telescope's schedule to accommodate observations of unexpected astronomical events or special tasks such as the Hubble Heritage Project.[10] The members of the group are quick to point out that they have used about twenty-five orbits annually, which is less than one percent of the total available observing time. By translating it into numeric terms, they hope to head off any criticism of this use of the telescope, a use that has as its primary purpose the creation of images for public consumption and not expressly for scientific research. Again, this caution reflects the uncertain status of pretty pictures within scientific institutions.

Although the Hubble Heritage Project carefully negotiates its position, the project demonstrates how much time, effort, and expense is dedicated to "pretty pictures." This seems to contradict the ambivalence inherent in the phrase, making its usage somewhat puzzling, and the ways in which the images are pretty requires clarification. For scientists, the label implies a comparison to numeric representations. Pictures cannot match the precision and elegance of equations or calculations, qualities scientists often associate with beauty. Also, the specificity of images closes off the possibility of making abstract and general statements about the nature of the cosmos, as is possible with statistical analysis.[11] The highest aesthetic regard, then, belongs to another mode of representation, and pretty pictures falter before scientists even evaluate their sensual qualities. By their very definition as pictures, they cannot attain the beauty of mathematics. Scientists also judge their images against other pictures. Here, too, they find the scientific images wanting because they do not match the aesthetic achievements of art. This, though, seems as much an acknowledgment of the expertise of artists, a modesty about the visual culture of science that recognizes a difference between an image made in the service of science and one made with more expressive intent. Although scientific images are made to be enjoyed, they remain rooted in science with its emphasis on knowing and communicating something about the physical phenomena and processes at work in the universe.

The pictures are pretty in still another way. Because of the conventions that govern what makes a scientific image attractive, especially the importance of composing and framing a coherent scene, pretty pictures rarely depict the regions that scientists find of greatest interest. While the Heritage images contain and convey scientific information, scientific advances happen more often at the frontier of representation, at the edge of resolution where an image dissolves into fuzz and blur. Scientists consider pictures with only these qualities to be unappealing or confusing to anyone who is not an expert in the field. As a result, pretty pictures depict the familiar objects that astronomers have known and observed for decades. The Hubble Space Telescope returns data that present that object in a new manner, but the Heritage images often depict favorite, iconic galaxies and nebulae. The images give scientists pleasure, often tinged with nostalgia because they remind them of their earliest experiences with a telescope or astronomy book, but they do not always lead astronomers to new insights.

Furthermore, while the Heritage Project images resonate with a wider audience, the promotional purposes have the potential to cheapen the images. They are made to sell science to those outside the small circle of experts, sullying the purity of the aesthetic response through their self-interestedness. Those who make pretty pictures avoid the phrase, describing them instead and almost interchangeably as attractive or compelling images. Both words grant them the power to spur further action—read a proposal, learn more about science, support research, inspire a career. Although this separates the images from the forgettable and momentary experience associated with prettiness, none of the actions call us to immerse ourselves further in the aesthetic experience. Instead, we are pushed toward the modes of representation that scientists value more highly.

It would seem no space remains for the sublime to enter the conversation about scientific images made for public viewing. In their aesthetic aspirations and their function within the practice of science, pretty pictures—even when refigured as compelling or attractive—seem firmly fixed in this world and not suggestive of the transcendence associated with the sublime. But this too quickly simplifies the nature of our response. The phrase "pretty pictures" refers to a category of images and is used across all scientific disciplines. As a result, it does not acknowledge or even admit the possibility of differences between individual examples. It makes no distinction between the mundane and the exotic, the accessible and the unreachable, the small and the vast. The heading lumps together an image of a yeast cell and a galaxy, assuming that they are both capable of producing a single and common response in the viewer. In its very definition, though, the sublime depends on the cosmos, and both Edmund Burke and Immanuel Kant cite it as a crucial source of the sublime experience. Violent storms, towering mountains, plunging abysses, and roiling oceans demonstrate nature's power and magnificence. While their grand scale contributes to their sublimity, these earthly phenomena combine size with terror, shock, and strength. But the immensity of the universe, its ability to express the infinite, is enough to evoke the sublime on its own, and the experience does not diminish through familiarity. In their subject matter, then, the Hubble Space Telescope images intersect with notions of the sublime in a manner unavailable to more mundane objects. Recognizing their scale and the large systems that link various phenomena—the human to a galaxy, a galaxy to the universe—opens up different aesthetic possibilities for these images.

Despite this, the impulse is to argue for the little guy, but not because yeast cells and galaxies are equal or they should be aligned with a grander aesthetic tradition, but because it points to another critical problem introduced by grouping all images under a single aesthetic heading. In a magnified view, a yeast cell—or some other tiny, everyday material—can become exotic, and it might, if presented in a certain manner, rival the monumentality of something much larger. Similarly, a poorly composed picture of a galaxy may fail to convey any sense of great size and scale, reducing the aesthetic impact of its representation. The means of representing the object are not given and fixed, and the choices made regarding color, contrast, and composition profoundly influence our aesthetic response. A summary judgment that all images made for display are pretty ignores the qualities evident in individual images.

To accurately assess the aesthetics of the Hubble images requires a closer look at specific examples, and the Heritage Project offers the prettiest and most aesthetically developed examples. The images still have all the lesser attributes that usually accompany pictures. They do not have the precise beauty of mathematics, nor do they necessarily show something unfamiliar or radically new. They do, however, emphasize the immensity of the cosmos, the great power of nature, and the insignificance of humanity—all attributes of the sublime. In the Keyhole Nebula, the light shines down on the clouds of gas, heightening the drama of the scene, and although this does record how they are illuminated, the extreme contrast depends on choices made during image processing to expand the range of tones and make detail more visible. The subtle texture, another aspect enhanced by careful adjustments to the contrast, gives the clouds three-dimensionality and volume, encouraging an analogy to rock formations. The orientation further encourages this comparison while also increasing the grandeur of the composition. North is at the bottom, and the choice creates a strong vertical composition with the brightest regions at the top. The light pulls our eye upward to intensify the scene's monumentality. Although we have no means to measure scale, appearance conveys a sense of incredible size. The colors are dynamic, and the yellowish orange clouds vibrate against the bluish green background. Considered on its aesthetic merits, it is not just pretty—it evokes the sublime.

The Hubble Heritage Project images repeatedly portray the cosmos as vast in scale and remind the viewer of humanity's relative unimportance. In an image of galaxy NGC 603, light again creates a dramatic effect. The dark edges contrast with the glowing regions in the center; we do not interpret the image as flat but see into the depths of another realm. The presence of a spiral galaxy on the lower left establishes a sense of the panoramic expanse represented. With only minimal knowledge of astronomy, we can recognize how diminutive the earth or our solar system would be. Other scenes similarly emphasize the enormity of the universe. The depiction of the Whirlpool Galaxy tightly frames its spiral structure in a vertigo-inducing composition. The arms that sweep from the top and bottom pull us into its center, which functions like a vanishing point. We are sucked into the apparent abyss of a swirling eddy. The gallery of images produced by the Hubble Heritage Project repeats these tropes of the sublime. When viewing these pictures, we cannot help but reflect on the vast size and scale of the universe and the majesty of the phenomena within it. Humanity's significance disappears, invisible in the face of the grandeur of the cosmos. As Howard Bond explains, the images show us that "vast things [are] going on in the Universe, which apparently have no regard to or are indifferent to us puny humans."[12]

The members of the Hubble Heritage Project repeatedly echo the themes of the sublime when they describe how they hope people will respond to the Hubble images. Zolt Levay wants the images to convey "the vastness of [the universe] and the varied nature of it . . . . It's huge things that we can't even comprehend how big they are and how violent some of these places are and at the same time, how beautiful it can be." Yet he also intends to communicate a sense of continuity, asking, "Can you really separate the landscape of the earth from a landscape on Mars or a landscape that's a million light years away? It's all the same thing." He goes on to suggest that recognizing the immensity of the cosmos and its unity gives rise to awe and wonder, and the experience can transport us outside the ordinary and everyday: "It would be wonderful to expand people's world view to include more than their day-to-day existence, to appreciate that we are a very small part of a much larger universe."[13]

Keith Noll similarly imagines that the Heritage images inspire dreams, hoping that "there are kids that have our pictures on their walls, who maybe spend some time dreaming about what it's like to be in space, what it would be like to travel to these exotic places." Although he begins with this basic goal of NASA's education programs—reaching kids and encouraging a lasting interest in science—his comments quickly turn in a more inclusive direction: "I don't care if all of them grow up to be scientists or not. Maybe some of them will. But what I really hope is that they all sort of carry this little bit of awe and mystery with them in their lives, so that everything isn't just about getting up and driving in traffic and paying the bills." Noll firmly believes that the Hubble images offer the possibility of transcending daily life, and although he starts by suggesting that the images can inspire children, the concerns he lists, traffic and bills, are much more adult. According to Noll, the Hubble images function as a reminder of "how amazing the world is, how amazing the universe is" and this is ultimately why he believes people like astronomy: it demonstrates our potential to reach beyond the apparent limits of daily existence.[14]

It would be tempting to stop here, to conclude that when we look more closely at individual examples and consider the intentions of their makers, we find that pretty pictures can be sublime. The longevity and popularity of Hubble images—they have not been dismissed, ignored, forgotten, or disregarded—support a claim for an aesthetic significance beyond prettiness. Their views of celestial phenomena remind us again and again of the power and vastness of the universe. But are they truly sublime? Such a judgment must also take into account how the images represent the universe, to consider more closely how the relationship between the pictures and the phenomena influences our aesthetic response. We could never see the galaxies, nebulae, and stars as they are portrayed in the Hubble images, not only because our vision does not equal the telescope, but also because the pictures themselves require human intervention. They depend on the imaginations of both those who produce them and those who see them. By adding color, changing the contrast, and composing the scenes, astronomers make visible aspects of the data that would be otherwise obscured. The results do not mirror the cosmos, but necessarily reinterpret upon it. Scientists would never mistake the images for mimetic pictures, but for the naive viewer, it is easy to imagine that the Hubble images offer a view of the universe we could experience if we had eyes that matched the telescope. It is with some disappointment that we recognize that they are mediated not only by an instrument but also by humans.

The Hubble images, then, might well be pretty as pictures, not in the comparative sense that usually accompanies the phrase, but in an ontological one. Their status as pictures, constructed visions of the universe, makes them less than we hope and opens them to questions. After all, Kant writes, "We must not point to the sublime in works of art, e.g. buildings, statues and like, where a human end determines the forms as well as the magnitude."[15] Instead, he argues that sublime arises only from an experience of raw nature. The Hubble images seemingly depict such raw, untouched nature, but ultimately such scenes become visible only through human intervention, through human hands that craft the images. If it were possible, a more direct representation without these needling doubts might elicit a more wholly sublime response. In the end, the Hubble images remind us of our ability to imagine and strive for such a perfect representation, and in that regard, they are pretty sublime.

Notes

1. Michael Lynch and Samuel Y. Edgerton, Jr., "Aesthetics and Digital Image Processing: Representational Craft in Contemporary Astronomy," in *Picturing Power: Visual Depiction and Social Relations*, ed. Gordon Fyfe and John Law, Sociological Review Monograph 35 (London: Routledge, 1988), pp. 184–221. Lynch and Edgerton's work is part of larger body of work by historians of both art and science on the uses and practices that adhere to images and other representations in science. A few representative examples include Peter Galison, *Image and Logic: A Material Culture of Microphysics* (Chicago: University of Chicago Press, 1997); Caroline Jones and Peter Galison, *Picturing Art, Producing Science* (New York: Routledge, 1998); James Elkins, *The Domain of Images* (Ithaca, NY: Cornell University Press, 1999); and Bruno Latour and Peter Weibel, *Iconoclash* (Cambridge, MA: MIT Press, 2002).

2. Elkins advocates for this in *The Domain of Images* and continues the project in *Visual Practices across the University* (Munich: Wilhelm Fink, 2007).

3. It also glosses over the possibility that scientists might gain insights about the data through the act of making a picture that effectively conveys information. Felice Frankel, herself a photographer who specializes in scientific images, argues for this in her work; see her *Envisioning Science: The Design and Craft for the Science Image* (Cambridge, MA: MIT Press, 2002).

4. For more detailed analysis of the different ways of representing the Whirlpool Galaxy, see Elizabeth A. Kessler, "Resolving the Nebulae: The Science and Art of Representing M51," *Studies in the History and Philosophy of Science* 38 (2007), pp. 477–491.

5. Anne Friedberg, *The Virtual Window from Alberti to Microsoft* (Cambridge, MA: MIT Press, 2006).

6. For discussion of earlier image-making practices in astronomy, see Alex Pang, "Technology, Aesthetics, and the Development of Astrophotography at the Lick Observatory," *Inscribing Science: Scientific Texts and the Materiality of Communication*, ed. T. Lenoir (Stanford, CA: Stanford University Press, 1998), pp. 223–248; and Simon Schaffer, "On Astronomical Drawing," in *Picturing Science, Producing Art*, ed. Caroline Jones and Peter Galison (New York: Routledge, 1998), pp. 441–474.

7. The democratization of image-making has resulted in publications that offer guidelines to scientists who are interested in producing their own images. See Travis Rector et al., "Image-Processing Techniques for the Creation of

Presentation-Quality Astronomical Images," *Astronomical Journal* 133 (2007), pp. 598–611; and Frankel, *Envisioning Science*.

8. Some of this history is retold by Keith Noll, "Hubble Heritage: From Inspiration to Realization" *STScI Newsletter* 18:3 (Fall 2001), pp. 1, 5.

9. For more information about the Hubble Heritage Project and its members, see their website (heritage.stsci.edu/index.html).

10. For example, this time was used to observe the collision of Comet Shoemaker-Levy 9 with Jupiter. The best known special projects are the Hubble Deep Fields, very long observations of small patches of sky that allowed astronomers to observe distant reaches of time and space.

11. Galison develops in detail the different values and uses associated with images and mathematical analysis in his study of microphysics, *Image and Logic*. To summarize and simplify, he argues that images are well suited to capturing "golden events," occurrences that prove the existence of something, while mathematics allow for the construction of general laws.

12. Howard Bond, September 16, 2003, oral history interview conducted by the author, Smithsonian National Air and Space Museum Archives, p. 5.

13. Zolt Levay, October 14, 2003, oral history interview conducted by the author, Smithsonian National Air and Space Museum Archives, p. 21.

14. Keith Noll, September 11, 2003, oral history interview conducted by the author, Smithsonian National Air and Space Museum Archives, p. 19.

15. Immanuel Kant, *Critique of Judgement*, trans. James Creed Meredith (Oxford: Oxford University Press, 2007), p. 83.

5

Against the Sublime

JAMES ELKINS

This essay is the result of a long interest in the sublime, which turned into a long dissatisfaction. I propose three different but connected arguments. First, the sublime is not well used as a transhistorical category: it does not apply outside particular ranges of artworks, most of them made in the nineteenth century. Second, in contemporary critical writing, the sublime is used principally as a way to smuggle covert religious meaning into texts that are putatively secular. Third, the postmodernist sublime is such an intricate concept that it is effectively useless without extensive qualification. In brief: saying something is sublime does not make it art or provide a judgment that can do much philosophical work or result in much understanding. I think the sublime needs to be abandoned as an interpretive tool, except in the cases of romantic and belated romantic art. Contemporary writers who use the word can always find synonyms to express what they mean, and those synonyms are apt to be more telling, and more useful, than the word "sublime."

The Sublime in Science

First a word about science. Most of my own entanglements with the sublime are recorded in my book *Six Stories from the End of Representation.*[1] It concerns images that are blurry, pixilated, dark, or otherwise inadequate to the objects they mean to represent. Many of the most interesting images of the period beginning around 1980, so I argue, are deeply concerned with the inadequacies and failures of representation. In astrophysics, for example, there are images taken at the limits of telescope's resolution, and in electron microscopy there are images of individual atoms. In painting and photography, the makers of the images took pleasure in those limits and did not try to improve upon them. Painters produced intentionally smeared canvases, and photographers made pictures deliberately out of focus. In the sciences, in contrast, those same kinds of images were not considered to be final or complete, and whenever possible the scientists improved their instrumentation to achieve clearer images. (Differences of intention were not my interest; I was concerned with the formal similarities: both the scientific images and the artworks were dark to the point of blackness, or so bright they were washed out, or blurred beyond recognition. It did not matter that the scientists tried to meliorate those qualities and that the artists tried to articulate them.)

My original notion for the book was to interpret all the images using the concept of the sublime. In the end I decided not to, because virtually none of the scientists used the concept in describing their own work. I sequestered the sublime to the two chapters dealing with fine art and used the scientists' own words to describe what happens in their images.

The sublime is inappropriate for talk about science for the simple reason that it is not in the vocabulary of most working scientists. I wanted to make sure that my chapters on astronomy, physics, and microscopy could be read by specialists in those fields, without a sense that their work was being misrepresented, or interpreted according to some master trope they had not known.

That is not to say that disciplines cannot be interpreted using language that the practitioners do not know (economics is a good example of a discipline that claims interpretive power over many fields whose practitioners could not understand its language), but to argue that it is sometimes more important to attend to the *particularities* of individual practices, their languages and even their equations, than to try to bind disciplines together using global languages. If the purpose is to understand a given scientific practice, there is a point when it becomes necessary to avoid interpretive agendas that are unknown to the makers and interpreters of the images in question.[2]

The Sublime in Visual Art

First argument: "the sublime" is a historically bounded term, not a transhistorical concept that can be applied to art or science in general.

The sublime, as elaborated by Kant, has a specific historical context, which makes it appropriate mainly for the interpretation of artworks in and around romanticism and its descendents. The postmodern sublime, first in Jean-François Lyotard's formulation and then in its many later forms (e.g., *Of the Sublime: Presence in Question*, edited by Jeffrey Librett; *Beauty and the Contemporary Sublime* by Jeremy Gilbert-Rolfe; and *The Sticky Sublime*, edited by Bill Beckley), is primarily an outgrowth of philosophy and literature and has no clear brief in the visual arts aside from individual authors' interests. The most prominent visually oriented formulation of the sublime, Mark C. Taylor's *Disfiguring: Art, Architecture, Religion*, has a clear application but to a very limited range of work and a specific set of interpretive possibilities. His remains a minority interest even regarding the artists (e.g., Anselm Kiefer), whom he privileges.[3] Aside from those moments, the Kantian sublime and the different postmodern sublimes run nearly exactly against the current of poststructuralist thinking, in that they posit a sense of presence and a nonverbal immediacy that short-circuits the principal interests of theorizing on art in the last thirty years, which are nearly all concerned with mediation, translation, deferral of meaning, miscommunication, and the social conditions of understanding. References to the sublime bypass all that literature—in some cases, it would be appropriate to identify that literature with poststructuralism in general—in favor of a direct access to pure, immediate presence. Very few authors—George Steiner might be the only widely visible example—have attempted to reinstate notions that depend on pure presence within a poststructural context. In scientific terms, it would be as if a group of artists were advocating a redescription of physics using only Newtonian models.

The Sublime In and Out of History

If I were to adopt an ahistorical stance, I would be able to speak more freely about the sublime. Presence would be easier to describe, for example, and so would transcendence. I could, in effect, count any image that points outside itself as sublime. A landscape painting points beyond itself simply by showing us objects that cannot be fixed—trees with swaying branches, clouds that move. An oil painting refers to things beyond itself with every gesture of the painter's brush. Following those widening gyres, the sublime could rapidly grow to encompass the entire history of images. It could even appear as if the sublime were the central problem of representation itself, as the philosopher Jean-Luc Nancy has suggested, or at least the crux of modern painting, as Lyotard says.[4]

A philosophical approach has its advantages, but it is open to historical objections. A short book by Tsang Lap-chuen, called *The Sublime*, can serve as an example.[5] His purpose is to find a theory of the sublime that is simply true, without any particular historical qualification. In the event, his theory reflects his reading: he knows Burke and Kant very well, Lyotard only slightly, and other recent writers not at all. He is entirely unacquainted with painting. Lap-chuen's brief description of Barnett Newman is prompted by his chance encounter with Lyotard's essay "Newman: The Instant," which he found after having been "occupied with the sublime for years." A number of recent authors have set out as Lap-chuen does, to "return," in another's words, "to the actual experience of the sublime."[6] But how can that make sense? What could it mean to define the sublime, once and for all, when it has changed so much since the first appearance of a word—later taken to be the same as the eighteenth-century sublime—in a classical text by Longinus? Lap-chuen's sublime is a specific sublime: one that belongs to a late-twentieth-century analytic philosopher who has only a passing interest in visual art. Historically, the possibility of applying the concept of sublimity to images is post-Kantian, because Kant himself was thinking of natural examples like the starry sky and the stormy ocean; the notion that sublimity is right for fine art and for science is largely confined to the second half of the twentieth century. That historical stricture is a bounding condition. When the sublime is applied without qualification (I mean, when there is no mention of *whose* sublime is being applied), then it is nearly always a matter of a simplified, received notion of Kant's sublime, and the implication is that the author is a Kantian of one sort or another. And that itself is problematic, even if it the author were to say, "I accept paragraphs 23 and 24 of the *Critique of Judgment*," because it has to be strange to be a faithful Kantian at the beginning of the twenty-first century, when so much has been written about Kant's analytic.

Still, the sublime is not purely a historical artifact. It is not a relic of the past, cut off from what seems true about pictures. More than other subjects, the sublime slips in and out of history in a bewildering fashion. Lyotard's lack of focus—his slurring of Henry Fuseli and Caspar David Friedrich, of Piet Mondrian and Gordon Onslow-Ford—is an intentional strategy: his sense of the sublime includes an awareness of its history *as well as* a conviction that it is an unavoidable element of experience. Part of the truth of the sublime, Lyotard might say, lies in the very broad tradition called "Western metaphysics" (that part can only appear true), and part lies in individual historical movements. Hence, some of his points are philosophical, and others are historical. The analytic philosopher Paul Crowther takes Lyotard to task for his lack of art historical precision: "To use the term 'sublime' to describe any artists who incline towards coloristic painterliness," Crowther says, "is so general as to be useless." On the other hand, it is hard to trust the historical judgments of a critic who can also claim that "the key artist in understanding the transition from modern to postmodern is Malcolm Morley."[7] (Morley is a photorealist.)

These slides in and out of history have been analyzed by Peter De Bolla in *The Discourse of the Sublime*. De Bolla distinguishes between a discourse *on* the sublime and a discourse *of* the sublime. The former includes texts that inquire into "the forms, causes, and effects of the sublime"—books on the subject of the sublime, we might say.[8] The authors of such books tend to cite "external authorities" and to divorce themselves from their analyses as far as possible. In the discourse *of* the sublime, authors produce sublime effects in their writing: the books themselves conjure and create the sublime. For the discourse on the sublime, the sublime effect is mostly out there, in the world; for the discourse of the sublime, it is found in "the interior mind." De Bolla says the two species are mostly divided by century: the discourse on the sublime is an eighteenth-century phenomenon, and the discourse of the sublime belongs to romanticism and the nineteenth century. They can also be seen as recurring moments in the sublime, and in that respect Lap-chuen's book is a discourse on the sublime, like Kant's *Critique of Judgment*. Burke's *Philosophical Enquiry into the Origin of Our Ideas of the Sublime and Beautiful* is partly a textbook on the sublime, and partly sublime itself—it produces sublime moments, it is *of* the sublime. De Bolla's own book is a Foucauldian analysis, and it is not a pure example of either of his own categories, but it is much closer in spirit to the earlier discourse on the sublime. De Bolla's voice is disengaged, many times removed from the close engagements of his authors. If it is possible to say "the sublime is an effect of the discursive analytic," then it will not be easy to conjure "sublime effects."

For an art historian or art critic, writing about the sublime is nearly always a matter of history—it is a discourse *on* the sublime. In art history, "sublimity" is known as a term applied in retrospect to Friedrich and a number of later romantic landscape painters in Germany and France, and it is a term found in art criticism beginning with the abstract expressionists. Applying the sublime to any other movements means taking increasing license with historical sources (as Lyotard, Crowther, Joseph Masheck, and many others have done by calling contemporary paintings "sublime"). It is historically inescapable that the sublime is a current critical term, and even if I would not go as far as Lyotard and claim that the sublime "may well be the single artistic sensibility to characterize the modern" or that aesthetics is "completely dominated" by the sublime, it is indispensable for any serious account of contemporary images simply because it is part of the current critical vocabulary.⁹ No matter how little sense references to the sublime may make, and no matter how little light they shed on the artworks that are said to embody them, the sublime is in the lexicon of contemporary art discourse. It would be artificial to exclude it altogether—as artificial as omitting such words as "representation," "realism," "image," and any number of other ill-defined terms.

One of De Bolla's central arguments is that the earlier discourse on the sublime was really a way of talking about subjectivity by keeping it at a safe distance. The writers' sense of their own subjectivities, De Bolla says, *were* the "unnamable," unrepresentable excess that they confidently assigned to the sublime. In effect the eighteenth-century philosophers hoped they could write in a controlled fashion about things that cannot be controlled. The nineteenth-century romantic discourse of the sublime came about when the pressures of that half-blinded way of writing broke the texts, and the writers' sense of their "unnamable" inner lives flooded into everything they wrote. In the eighteenth century, writers kept their texts (if not their minds) free of the dangerous influx of subjectivity by shoring up their writing with theories. Theology, ethics, physiognomics, logic, and even optics were all brought to bear on the sublime, as if it needed help from professionals in other fields in order to remain coherent. It is a signal virtue of De Bolla's rather cold account that he can explain this odd frame of mind: it is an effect *of* the sublime itself, clear evidence that the sublime cannot be adequately explored unless the writing finds a way to move back and forth from discourse *on* to discourse *of*.

Is it possible to say how much of the sublime is ours (part of history, something we can write *on*) and how much is us (part of experience, something that can only be written *of*)? There is some evidence that the sublime still owns us and that writers cannot do much more than measure and describe it. The critic Suzanne Guerlac has argued that the sublime is the unnamed theory that framed theory itself during the inception of deconstruction and French semiotics between 1970 and 1974. The sublime, she says, "enabled the constitution of theory as a subject" for Julia Kristeva, Oswald Ducrot, and Tzvetan Todorov, giving their enterprise its model of transgression. Guerlac wrote her essay in 1991, almost twenty years after the events she describes, but even then she thought the sublime was "still not an object of theory" but a set of conditions *for* theory.[10] I do not find Guerlac's argument entirely convincing, because the sublime lends itself easily to all kinds of "limit situations," transgressions, and failures of theory, but it is certainly true that the models of radical thought embodied in Kristeva's *Révolution du langage poétique*, one of the founding texts of poststructuralism, can be described with suspicious ease as models of Kant's mathematical sublime. At the least it is clear that some poststructuralist thinking owes the sublime a debt that it still cannot quite acknowledge.

All this is to say that the sublime is not susceptible to a full analysis: it still produces its effects, and it still entangles many contemporary ideas from representation to transcendence. The more subtle a theory of the sublime becomes, the more the sublime threatens to overcome it. The critic who has pushed hardest on the sublime, Paul de Man, moved through a series of increasingly hyperbolic theories to the position that language itself is inhuman and that any creation of human meaning is a form of "madness."[11] In that account, the sublime can be read as a model of meaning itself: the willed, inescapable imposition of sense onto an incommensurate substrate of language, and the ensuing game of reading and awareness of reading. Reading de Man on the sublime is like being half-drowned: every page is waterlogged with his painful awareness of the impossibility of ending the game.

The critic Martin Donougho is right to wonder how seriously Kant, Thomas Weiskel, de Man, De Bolla, and Steven Knapp take the sublime, and whether—or how—they believe in it.[12] In that way De Bolla's literary, or Foucauldian, sense of writing *on* and *of* becomes, more directly, a question of *belief*.

Why the Sublime Is a Religious Concept

There is another, possibly deeper reason why the sublime matters to a contemporary sense of pictures, and why it is so important—and so vexed, and often so opaque—in literary theory. Talking about the sublime is a way of addressing something that can no longer be called by any of its traditional names, something so important that words like "art" would be crippled without it: the possibility of a truth beyond the world of experience (and not merely beyond the world of articulation, or representation).

In past centuries, some of the ideas now contested under the name "sublime" were known more directly as religious truth or revelation. Today, writers in the humanities mostly shy away from open talk about religion. Such words as "sublimity," "transcendence," and "presence," shrouded in clouds of secular criticism, serve to suggest religious meanings without making them explicit. For many reasons, the sublime has come to be the place where thoughts about religious truth, revelation, and other more or less unusable concepts have congregated. An example that is often cited is from Weiskel's influential book *The Romantic Sublime: Studies in the Structure and Psychology of Transcendence* (with the word "transcendence" taken in a philosophical sense). Weiskel says, all at once (and only once) that "the essential claim of the sublime is that man can, in feeling or speech, transcend the human." A "'humanist sublime,'" Weiskel thinks, "is an oxymoron," because the sublime "founders" without "some notion of the beyond." At the same time, he will not write about the religious aspect of his subject. He closes the subject peremptorily: "What, if anything, lies beyond the human—God or the gods, the dæmon or Nature—is matter for great disagreement."[13] That is on page 3; afterward, he keeps quiet about religious meaning.

I would not be quite as silent about religious meanings as Weiskel, but it is a matter of knowing when to speak and how much to say. For example, it is important not to assume that "the sublime," "presence," and "transcendence" are philosophical masks that can be removed, revealing a hidden religious discourse. They *are* that discourse: they are taken by authors like Weiskel to be the only remaining ways in which truths that used to be called religious can find voice within much of contemporary thought. In one sense the dozens of twentieth-century books that discuss the sublime are interrogating the possibility of religious experience, but in another sense—the only one available for reflective writing—they are not addressing religion but asking only about the coherence and usefulness of the sublime.

This permeable veil between two kinds of thinking has always been a trait of the sublime. Kant is adamant about the separation (he protests too much), Longinus talks uncertainly of divinity, and Weiskel permits himself the one apostrophe. The same veil comes down in front of religious writers when they look across at the sublime from theology and religious history. Rudolf Otto's book *The Idea of the Holy*—famous partly because it introduces the wonderful word "numinous"—skirts the sublime, as if Otto is unsure whether the sublime is part of the holy. His book is one of the best to study the vacillation about the sublime and religious writing. At one point he says the sublime is a "pale reflection" of numinous revelation; five pages later he says the sublime is "an authentic scheme of 'the holy.'" Otto is a neo-Kantian, and he sees

Kant's aesthetics as unmoored talk about the holy. There is a "hidden kinship," Otto concludes, "between the numinous and the sublime which is something more than a merely accidental analogy, and to which Kant's *Critique of Judgment* bears distant witness."[14] What an odd phrase, "something more than a merely accidental analogy," especially in a book devoted entirely to a *systematic* reappraisal of Kant: it is entirely typical of the tenuous alienation that still obtains between religious vocabulary and the sublime.

Religious Concepts in Art Criticism and History

Michael Fried's "Art and Objecthood" may be the most frequently cited example of religious tropes in art history. A standard reading is that Fried's formalism in that essay is one of the few channels remaining for religious discourse in the almost wholly secular domain of modernist criticism. Such a reading, however, is a way of talking about religion and modern or postmodern art in the same text. Any project that disregards the dynamic, and seeks only to illuminate the religious within the nonreligious, is missing the pressure—*historical* pressure, Fried would say—to *not* speak directly of the religious in the context of modernism. Although I cannot do more in this essay, I would suggest that to move forward, contemporary art criticism might begin by acknowledging that the sublime cannot be fully excavated from its crypto-religious contexts.

Poor Sublime

Some words, then, about the weakness of the concept. The sublime has been roundly critiqued by a number of writers for its direct appeal to pure presence, and its obliviousness to poststructural doubts. It has also been criticized because it leads scholars (like me!) to focus on images of things that are incomprehensibly vast, or unimaginably small, or frighteningly blank, dark, blurred, smeared, pixilated, or otherwise illegible. The sublime, so it is said, takes people away from the real world of politics and society, of meaning and narrative, of culture and value.

Poor anemic sublime. Poor elitist concept, born in the leisured classes of eighteenth-century Europe, lingering on into the twenty-first century as an academic hothouse plant. "One should see the quest for the sublime," according to the philosopher Richard Rorty, "as one of the prettier unforced blue flowers of bourgeois culture."[15] (He says that the sublime is "wildly irrelevant to the attempt at communicative consensus which is the vital force" of common culture.)

Poor sublime, in that case, which can only express the most atrophied and delicate emotions of distance and nostalgia, which requires a battery of arcane ideas to keep it afloat, which can only be found in the most hermetic postmodern art or the most *recherché* (old-fashioned) romantic painting. Poor sublime, which can only sing a feeble plaintive song about longing, which has nothing to say about the things that count in visual culture—especially gender, identity, and politics. Poor irrelevant sublime, as Philippe Lacoue-Labarthe says, which "forms a minor tradition," following along after beauty is exhausted.[16]

Poor sublime, too, which seems like "a thoroughly ideological category" (as Terry Eagleton says), or a "discourse" with certain "effects" that need dispassionate Foucauldian study (as De Bolla proposes).[17] *Irresponsible* sublime, which puts all kinds of things beyond the reach of critical thought, and so "becomes the luxury of the aesthete all over again," protecting postmodernists from having to make difficult judgments.[18] Pallid sublime, which mingles with beauty and makes for easy pleasures.

Poor sublime: relic of other centuries, perennially misused as an attractive way to express the power of art, kept afloat by academics interested in other people's ideas, used—ineffectually, I have argued—as a covertly religious term, to permit academics to speak about religion while remaining appropriately secular. And finally, poor sublime, exiled from contemporary philosophy even as it suffuses so much of it. [19]

In the end, the sublime is damaged goods. It has been asked to do too much work for too many reasons, and it has become weak. I propose a moratorium on the word: let us say what we admire in art and science, but let us say it directly, using words that are fresh and exact. Sometimes it is impossible to avoid such words as "awe," "wonder," and "the sublime," but they are so loaded with history, and at the same time so vague, that they are nearly impossible to control—they are like enormous cargo ships that need to be steered very carefully and slowly.

So, what to do? I have only one small proposal along these lines, which is not philosophical and does not solve the problem but has the virtue of being practical and, potentially at least, producing interesting writing. As novelists know, it is always possible to describe an experience instead of assigning it to a generic term. As novelists know, it is often better to give a concrete example of an action instead of naming the emotion that provoked the action. "He threw the porcelain plate onto the slate floor" reads better than "He was very irritated." The same strategy—preferring the particular, focusing on the event rather than the emotion—can convey many nuances of the sublime. What do I say when I am ambushed by the tremendous appearance of the Milky Way, pouring from one horizon to another, with Cygnus gleaming in its middle, as I was this summer in the west of Ireland? I prefer to try to say it as I see it, in words as sharp as I can manage, avoiding words like "awe," "wonder," and "the sublime"—words that have been said without thought so many times that they are like blank coins, rubbed by thousands of fingers until they are nothing but thin blank disks.

Notes

1. James Elkins, *Six Stories from the End of Representation: Images in Painting, Photography, Microscopy, Astronomy, Particle Physics, and Quantum Mechanics, 1980–2000* (Palo Alto, CA: Stanford University Press, 2007). This essay is an abbreviated version of one I am preparing for a book on critical terms in art theory.

2. On the vexed tradition of importing concepts from art to interpret science, see Elkins, *Six Stories*, and Elkins (ed.), *Visual Practices across the University* (Munich: Wilhelm Fink, 2007).

3. See Mark C. Taylor, *Disfiguring: Art, Architecture, Religion* (Chicago: University of Chicago Press, 1992).

4. Jean-Luc Nancy, "The Sublime Offering," in *Of the Sublime: Presence in Question*, ed. Jeffrey Librett (Albany: State University of New York Press, 1993), pp. 25–54, at p. 50; see also "Preface to the French Edition," pp. 1–3, especially p. 1.

5. Tsang Lap-chuen, *The Sublime: Groundwork Towards a Theory* (Rochester, NY: University of Rochester Press, 1998).

6. Richard White, "The Sublime and the Other," *Heythrop Journal* 38 (1997): 125–143, at p. 125.

7. Paul Crowther, *Critical Aesthetics and Postmodernism* (Oxford: Clarendon Press, 1993), pp. 159, 187.

8. Peter De Bolla, *The Discourse of the Sublime: Readings in History, Aesthetics, and the Subject* (New York: Oxford University Press, 1989), p. 30.

9. Lyotard, "Response to Philippe Lacoue-Labarthe," trans. Geoff Bennington, in *Postmodernism, ICA Documents*, ed. Lisa Appignanesi (London: Free Association Books, 1989), p. 15.

10. Suzanne Guerlac, "The Sublime in Theory," *MLN* 106 (1991): 895–909, at pp. 895 and 909.

11. Paul de Man, *Rhetoric of Romanticism* (New York: Columbia University Press, 1984), p. 122.

12. Martin Donougho, "Stages of the Sublime in North America," *MLN* 115 no. 5 (December 2000): 909–40.

13. Thomas Weiskel, *The Romantic Sublime: Studies in the Structure and Psychology of Transcendence* (Baltimore, MD: Johns Hopkins University Press, 1976), p. 3.

14. Rudolf Otto, *The Idea of the Holy: An Inquiry into the Non-Rational Factor in the Idea of the Divine and Its Relation to the Rational* (1917), trans. John Harvey (Oxford: Oxford University Press, 1946), pp. 42, 47, 65.

15. Richard Rorty, "Habermas and Lyotard on Modernity," in *Habermas and Modernity*, ed. Richard Bernstein (Cambridge: Polity Press, 1985), pp. 161–175, at p. 174.

16. Philippe Lacoue-Labarthe, "Sublime Truth," in *Of the Sublime*, pp. 71–108, at p. 84.

17. Terry Eagleton, *The Ideology of the Aesthetic* (Oxford: Blackwell, 1990), p. 90; De Bolla, *Discourse of the Sublime*, p. 35.

18. This is part of Timothy Engström's trenchant critique of Lyotard's sublime, which he says "runs the risk of putting beyond narrative, beyond critical and discursive reach, the sorts of pains and excesses that narratives produce ... a death camp here, the odd effort at genocide there." "The Postmodern Sublime?: Philosophical Rehabilitations and Pragmatic Evasions," *boundary 2* 20 no. 2 (1993), p. 194.

19. Although philosophy per se is not this subject of this essay, it is especially damaging to the coherence or usefulness of the sublime that it is so malleable that it can come to stand for poststructuralist theory itself, as Guerlac argues and Nancy implies.

6

Neuroscience and the Sublime in Art and Science

JOHN ONIANS

Is it possible to apply neuroscience to the sublime? Not without reducing it to the ridiculous, some would say. Others would disagree, beginning with Edmund Burke, who did more than anyone to bring the sublime to the center of discussions of human aesthetic response, preparing the ground for Kant's more philosophical reflections. Burke strongly believed that many human responses, such as that to the sublime, have their roots in our biology, and when he says that many feelings, such as sympathy, "arise from the mechanical structure of our bodies . . . or from the natural frame and constitution of our minds,"[1] he is certainly referring to what we would now call our neural constitution. He would have welcomed the fact that neuroscience has developed so rapidly in recent decades that we can use it to explore the sublime.

Things have been moving in his direction for some time. Burke's view was based only on his observation of human behavior and his following up of linguistic clues that led him to his conclusions. The universality of the sublime, for example, was demonstrated by the presence in all languages of words for similar related emotions:

> Several languages . . . use the same word to signify indifferently the modes of astonishment or admiration and those of terror.

> *Thambos* is in Greek either fear or wonder .... The Romans used the verb *stupeo* a term which strongly makes the state of astonishment ... and do not the French *étonnement* and the English *astonishment* and *amazement* point out the kindred emotions which attend fear and wonder? They who have a more general knowledge of languages could produce, I make no doubt, many other and equally striking examples.[2]

All peoples share the core response to the sublime, a mixture of admiration and terror. Burke also realized that this response was so universal and so powerful because it was crucially connected to "self-preservation." It is because the passions, "which are conversant about the preservation of the individual, turn on *pain* and *danger*" that "whatever is fitted in any sort to exercise the ideas of pain, and danger, that is whatever is in any sort terrible, or is conversant about terrible objects, or operates in a manner analogous to terror is a source of the *sublime*; that is it is productive of the strongest emotion which the mind is capable of feeling."[3]

The link between the intensity of such emotions and the drive for survival was taken up by Charles Darwin, who had visited a wide range of different peoples during his voyage on the *Beagle* and had observed human, as well as animal, reactions in all sorts of circumstances. These researches allowed him, in *The Expression of the Emotions in Man and Animals* (1872), to begin to pin down the common physiological and psychological basis of some critical responses. He understood the behaviors associated with them, could identify the muscles involved, and recognized that they were ultimately controlled by the nerves: "Certain actions, which we recognise as expressive of certain states of mind, are the direct result of the constitution of the nervous system."[4] He also understood that we share key features of these responses not just with other primates:

With all, or almost all animals, even with birds, Terror causes the body to tremble. The skin becomes pale, sweat breaks out, and the hair bristles. The secretions of the alimentary canal and of the kidneys are increased, and they are involuntarily voided, owing to the relaxation of the sphincter muscles, as is known to be the case with man, and as I have seen with cattle, dogs, cats and monkeys.[5]

Today, understanding in both areas has greatly advanced. Half a century ago ethologists, such as Konrad Lorenz and Nikko Tinbergen, showed the importance of emotional responses throughout the animal kingdom, and later Desmond Morris, in books like *The Naked Ape* (1967) and *The Pocket Guide to Manwatching* (1982), went on to explore how they are expressed in human relationships in ways that are both inborn and acquired. Neuroscientists had, already in the nineteenth century, begun to identify the separate areas of the brain involved in receiving information from the senses and in generating motor outputs. In the twentieth century they started to understand the process by which they were linked in such a way that a particular visual experience might lead to a particular motor action. It was, however, only in the last decades of the century, and especially with the development of a brain scanning technique known as functional magnetic resonance imaging (fMRI), which monitors the brain's activity in real time by tracking changes in blood flow, that it became possible to begin to identify the neural correlates of particular experiences. We are still a long way from being able to explain the detail of the neuropsychology involved, but some of the principles that underpin it are now becoming clear. This is especially true in relation to responses that are triggered by visual experience, due especially to the work of Semir Zeki, whose book *Inner Vision* appeared in 1999. Modern neuroscience allows us at last to begin to flesh out Burke's notion of "the natural frame and constitution of our minds" as it relates to the sublime.

But, before we attempt to do this, we first need to briefly analyze the sublime, clarifying the core ideas with which it has been associated over the two millennia that it has been current. Probably the most important is the notion of great scale and the response it evokes. Certainly it is metaphorical "height" or "grandeur" of rhetorical style that is the concern of the treatise *Peri Hypsous* (*On Height*) usually translated in Latin as *De Sublimitate*, once incorrectly attributed to the first-century rhetorician Cassius Longinus. For Burke, who was the first to explore "the sublime" in a much wider range of contexts, especially nature and art as well as literature, there are many other important elements, including, besides "vastness," "obscurity," "power," "infinity," and "difficulty," all of which are contrasted with the attributes of "the beautiful."

Kant took up Burke's pairing already in an early essay, *Observations on the Feeling of the Beautiful and the Sublime* (1764), but it was in the *Critique of Judgement*, first published in 1790, that he developed his ideas in ways that moved discussion of the sublime to a new level. As he says, he replaced Burke's "physiological" framework, with its emphasis on "self-preservation," with one that was "transcendental."[6] This completely changes the status of his great "analytic of the sublime" that occupies the second book, being treated at nearly three times the length of the "analytic of the beautiful" covered in the first book. The chief reason for this disparity is that for Kant the experience of the beautiful is associated with "freedom" and "play," while the sublime demands serious mental effort. There are many ways of differentiating this effort. One is by his claim that while "the beautiful" is involved with "quality," "the sublime" is involved with "quantity." More important, though, is the point that the very essence of the sublime consists in recognizing that the sublime is located not in the

FIGURE 5 Mother and Child

Courtesy of Justine Cooper

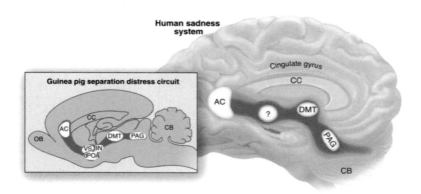

FIGURE 6 There are remarkable similarities between regions of the guinea pig brain that when activated provoke separation distress, and areas of the human brain that are activated during feelings of sadness. The Anterior Cingulate (AC) is critically important for relating basic social emotions to cognitive decision making. The basic circuitry for separation distress is highly concentrated in ventral septum (VS) dorsal Preoptic Area (dPOA) and the bed nucleus of the stria terminalis (BN), all necessary for separation distress interacting with many other social processes such as sexuality, maternal nurturance, as well as other emotions. OB is the olfactory bulb; CC, corpus callosum; CB, cerebellum. The basic circuitry for the psychic pain of social loss includes the dorsomedial thalamic (DMT) and midbrain periaqueductal gray (PAG) areas. To the best of our knowledge, the activity of this whole complex network constitutes a fundamental substrate for the affective experience of sadness in humans. The illustration is from J. Panksepp, *Science*, 302, (2003): 237–239.

FIGURE 7 Ferenczy Károly, *Archaeology*, 1896

Courtesy of the Hungarian National Gallery

FIGURE 8 Keyhole Nebula

NASA and The Hubble Heritage Team (STScI/AURA)

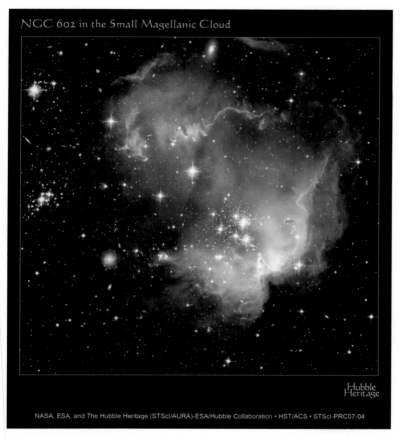

FIGURE 9 NGC 602

NASA, ESA, and the Hubble Heritage Team (STScI/AURA)-ESA/Hubble Collaboration

FIGURE 10 Whirlpool Galaxy (M51)

NASA and The Hubble Heritage Team (STScI/AURA)

FIGURE 11 Automata in Geppetto's workshop, from *Pinocchio*, 1940
© *Disney Enterprises, Inc.*

FIGURE 12 Pinocchio and Blue Fairy, from *Pinocchio*, 1940

© *Disney Enterprises, Inc.*

FIGURE 13 The sublimation of iodine

Photo courtesy of Rodney Schreiner (University of Wisconsin)

FIGURE 14 Paul Klee, *Eros*, 1923

Courtesy of Rosengart Collection, Museum Lucerne

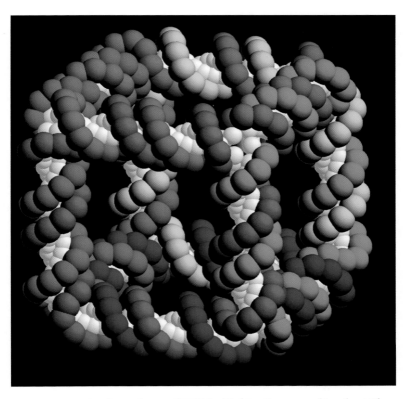

FIGURE 15 A cube made out of DNA by Nadrian Seeman and Junghuei Chen

Courtesy of Nadrian Seeman and Junghuei Chen

object but in the mental activity it stimulated. "The concept of the sublime in nature" thus "gives on the whole no indication of anything final in nature itself, but only in the possible *employment* of our intuitions of it in inducing a feeling in our own selves of a finality quite independent of nature."[7]

Kant's analysis of this mental activity in turn provided the starting point for discussions of the sublime that continue until today, as in the writings of Jean-François Lyotard, who used the category of the sublime to reflect on the limits of rational understanding and the postmodern condition. These discussions embrace the sublime in nature, art, literature, music, and other fields, as they already did for Burke. They also constantly refer back to his key associations, greatness, height, obscurity, and so on, while greatly expanding on them. Today, the sublime might be invoked in the context of the astronomer's black holes, a Mark Rothko painting, a Jorge Luis Borges novel, a Karlheinz Stockhausen piano piece, or any work that challenges us emotionally and intellectually.

The most basic issue that neuroscience helps us to understand is Burke's association of the sublime with self-preservation. When he made the link, he was simply trying to explain why we have such strong emotions about large and dangerous things. He did not know, as we do now, that this is the case because of the operation of natural selection on our neural apparatus. There are many aspects to this, but two may concern us here—survival and the transmission of our genetic material, because they help us understand why we react to large things and why we react to them so strongly. The most fundamental point is that because those individuals who had strong emotional reactions to large and thus dangerous things were more likely to survive than those who did not; such individuals were also more likely to transmit the genetic material that coded for the neural structures that support those reactions. That is why those reactions are inborn as part of our makeup and why they are so strong. Two stages in the reaction are critical. The first is the discrimination of a large object, and the second is the strength and rapidity of the response. Each has a specific neural basis.

The discrimination of large objects is sustained by particular resources in the area of the brain that deals with visual information coming from the retina. These include "feature detectors," arrays of different types of neurons, each of which reacts to a line of a different orientation, which allow us to perceive the edges of objects in the visual field and so identify them as "gestalts," or separate shapes. They also, in conjunction with other areas, help us to estimate scale. They allow us to identify a large object in our environment, and, in us, as in all animals, they cause us to pay particular attention to that object, because it might be a large member of our own or another species and so liable to be either very helpful or very dangerous. We are born with neural resources that encourage us to do this, which is why a baby will pay particular attention to a shape looming over it. A baby might expect the shape to be a beneficial parent, but as we grow up our neural resources, including those of memory, ensure that we become wary of such large shapes, as inborn fears are confirmed by our experience of the power first of other adults who are less kind to us and then of larger rivals. Neuroscience teaches us that the resources that sustain vision also sustain the imagination, which is why, as Burke reminds us, we are so intimidated by the greatest thing imaginable—God, who is "invested upon every side with omnipresence," so "we shrink into the minuteness of our nature and are, in a manner, annihilated before him."[8] God's superlative greatness gives him superlative power, and that fills us with dread and respect.

We need not only to discriminate the large and powerful but also to immediately be prepared to respond to them appropriately, and this we do because of another element in our neural makeup: chemical neurotransmitters. In circumstances where self-preservation is at issue, one of the most important of these is norepinephrine. This is a substance that is distributed through neurons from a single center at the base of the brain, the locus coeruleus, to those parts whose activation is necessary in such conditions. If it was not distributed so quickly and effectively through the nervous system, we would be much less likely to survive dangerous encounters. Equally important is its impact on the body once distributed. It increases the heart rate, triggers the release of glucose from energy stores, and improves muscle readiness. And these are particularly telling manifestations because they help us to understand one of the most intriguing aspects of our strong negative emotional response to a life-threatening situation: the sense that we can seem to enjoy it. Thus, several of these reactions are ones that make us feel more alert and engaged and so make us feel good. This is why the experience of the sublime may be one we seek and, when we obtain it, that we seek to prolong. The early devotees of the sublime would have felt this when standing on a mountaintop, and a modern mountain climber feels it even more intensely when spread-eagled on a vertical face. Burke is clear that pleasure can come out of pain, and in order to distinguish it from a pleasure, which has a positive source, he calls it delight. "The delight which arises from the modification of pain, confesses the stock from whence it sprung, in its solid, strong and severe nature."[9] That is a pleasure a mountain climber might recognize.

Neuroscience helps us to understand Burke's "physiological" sublime, and it also gives us insights into the more elusive and "transcendental" sublime that interested Kant. There are many aspects to Kant's transcendental sublime, but one of the most striking is the notion that it is not the natural object, the mountain or ocean, that is sublime, but the human response to it, what he defines as the "*employment* of our intuitions of it in inducing a feeling in our own selves of a finality quite independent of nature."

For the feeling of the sublime involves as its characteristic feature a mental *movement* combined with the estimate of the object, whereas taste in respect to the beautiful presupposes that the mind is in *restful* contemplation and preserves it in this state. But this movement has to be estimated as subjectively final (since the sublime pleases). Hence it is referred through the imagination either to the faculty of *cognition* or to that of *desire*, but to whichever faculty the reference is made, the finality of the given presentation is estimated only in respect of those faculties (apart from end or interest). Accordingly, the first is attributed to the object as a mathematical, the second as a dynamical, affection of the imagination. Hence, we get the above double mode of representing the object as sublime.[10]

Kant's explanation is far from transparent, but one way it can be related to neuroscience is suggested by current views of the brain, especially as they relate to the processing of visual experience. The part of the brain at the rear of the skull concerned with vision, the place where the nerves leading from the retina terminate, does much more than facilitate the discrimination of large objects, as noted earlier. It plays a critical role in coordinating all our relationships with the world. Essentially, information from the retina passes through a series of areas, among which V1, V2, V3, V4, and V5 are the best understood, before being processed by other areas. Part of that processing involved some information passing to a so-called "ventral stream," going downward to the side of the brain, and a "dorsal stream" going upward. The role of the ventral stream, which leads to the temporal lobes, is principally to identify the objects in our environment. The role of the dorsal stream is to determine where they are and prepare us to take actions appropriate to them.

There is in this a surprising correspondence with Kant's framework. The visual region in general plays exactly the role Kant envisages for the imagination, since it is where the first mental representation takes place, and not just the mental representation of visual information. One of the most interesting of recent discoveries is that the visual area is activated by words as well as visual signals, with the word "red," for example, activating the area of V4 that is concerned with color and with whole scenes being imagined using the resources of all the visual areas. This is why the visual brain can be seen as the theater of representation par excellence, the place where words are turned into images. Again, the ventral stream, being concerned with object recognition, is occupied precisely with the task of *cognition*, while the dorsal stream, which is where the location of objects is determined and movements in relation to them are prepared for, is precisely occupied with matters relating to *desire*. Not only do Kant's separate faculties have correlates in the brain, but the brain also recapitulates the relation between them. And, given that that relation requires constant give-and-take and feedback, it is not inappropriate to see all collectively as involved in what he calls *mental movement*.

We can even go on to analyze in some detail the way the particular *mental movement* associated with sublime objects is aroused. Kant makes clear that it is not the ocean as such that creates such mental movement but the ocean when in a state of chaos, because it then becomes undefinable and confused, and this observation recalls some of the findings of some of the most remarkable experiments of Zeki. Already in 1998 Zeki published experiments, redescribed in *Inner Vision*, showing that there is a substantial difference in the brain's response to normally and abnormally colored objects.[11] When we look at normally colored objects, the areas that are activated are, besides V4 that is concerned with color, the area toward the temporal lobe that is concerned with object recognition and the hippocampus, which plays an important part in memory, as well as the inferior frontal convolution of the right hemisphere. In other words, we can observe the brain doing what it is designed to do: noting the object's color, identifying it by comparison to objects it has seen before, and remembering what its relation was to such objects. When the eye is presented with unnaturally colored objects, V4 is again activated, but there is no visible hippocampal activity, and a different frontal area is involved: the middle frontal gyrus. This an area known to play a critical role in making judgments about the appropriateness of an action, often known as a "monitoring station," and it is easy to see how, given the indeterminacy of the object, such judgments become particularly difficult. This picture has been further elaborated by subsequent experiments. These showed that, when looking a normally colored object, such as a red strawberry, the visual brain readily establishes a stable connection with the object recognition area in the temporal lobe, but when an object such as a blue strawberry is presented to the eye, brain activity spreads to other areas and moves around, as if in search of some resolution of the problem that it is constantly setting itself: "what is this object?" and "what must I do with it?" From all of this, it is clear that, while an object with clear positive associations, such as a red strawberry, elicits a response that is essentially *restful* and positive, an indeterminate or confused object is indeed more likely to provoke mental movement.

This is not to argue that there is a precise correlation between Kant's view of the mental movements associated with the sublime and the neural activity revealed by Zeki's scans, any more than that Kant had a prescient grasp of the principles of neuroscience, but the correlation between Kant's views and Zeki's experimental observations suggests that Kant was indeed able, using only his own acute powers of analysis, to come to conclusions about the way the mind works that do anticipate in some ways the findings of modern brain research. Zeki claims more than once that the works of some painters relate so closely to the findings of neuroscience that it is possible to consider them to be neurologists. He also implies that the same is true of Kant and other philosophers. The argument here points to a similar implicit conclusion.

So much for the theory of the relationship between some aspects of the sublime, as dealt with by Burke and Kant, and neuroscience. What about the possibility of using neuroscience to develop a system of practical criticism of the sublime? What follows here is a sketch of such a system. It is inspired by a concert I attended recently in Fischer von Erlach's Kollegienkirche (1694) in Salzburg, a performance of a piece by the French avant-garde composer Gérard Grisey (1946–1998). The title of the piece itself evoked the sublime "Le Noir de l'Etoile (1989/90) pour six percussionistes disposés autour du public, bande magnetique et transmission in situ de signaux astronomiques," ['Blackness of the star (1989/90) for six percussionists arranged around the audience, magnetic tape and on-site transmission of astronomic signals'] and the setting certainly meets some of the criteria for the sublime proposed by both Burke and Kant. The first is that of absolute size, since the church is one of the largest and grandest designed by Fischer von Erlach. Besides, with its four symmetrical arms ending in curved apses and crowned by massive vaults supported by gigantic columns, it meets several, though not all, of the particular requirements of Burke's section on "Succession and Uniformity." "Succession and uniformity of parts are what constitute the artificial infinite,"[12] succession because "frequent impulses on the sense . . . impress the imagination with an idea of their progress beyond the actual limits," and uniformity because this is best achieved by a "rotund" configuration, because "turn which way you will, the same object seems to continue and the imagination has no rest."[13] In accordance with Burke's requirement for "obscurity," the performance was also in complete darkness, except for the six musicians, separately illuminated, two at each end of the nave and one in each of the short transepts. "To make anything very terrible, obscurity seems in general to be necessary. When we know the full extent of any danger, when we can accustom our eyes to it, a great deal of the apprehension vanishes."[14] Burke goes on to note that this is the atmosphere in which notions of ghosts and goblins affect our minds, an idea that chimed with Grisey's description of himself as a member of the school of "spectral" music. The music, consisting of a constant exchange between the six vast batteries of tympani, met another of Burke's criteria, spelled out in the section on "Sound and Loudness": "The eye is not the only organ of sensation, by which a sublime passion may be produced. Sounds have a great power in these as in other passions . . . Excessive loudness alone is sufficient to overpower the soul, to suspend its action, and to fill it with terror."[15]

Yet other requirements of the sublime were met by the music's introduction. This began with a disembodied voice coming from a loudspeaker high up toward the vault, telling the audience how the composition was inspired by the notion of extinct supernovas called pulsars, and Grisey's awareness of the way the pulses of energy they emitted could be received and transformed into sound by a radio telescope. The emphasis placed on the unimaginable distance of such pulsars from us in space and time and the extraordinary density of their mass placed the event securely in the domain of Kant's "mathematically sublime."[16] For those who could follow it, the account met another specific requirement of his: "The sublime is that, the mere capacity of thinking which evidences a faculty of mind transcending every standard of sense."[17]

All of this demonstrates that the event had elements of the sublime. There were, however, aspects that reduced this sublimity.

These included the light on the tympanists, which diminished the "obscurity," and the church's cruciform shape, which, according to Burke, necessarily diminishes the effect of "uniformity." But perhaps the most interesting reduction in sublimity came from the modifications made to the space of the original church in the name of improved acoustics. On the day before the concert, I had heard a lecture from the acoustics engineer involved, Karlheinz Müller. He had told us of the conflict between those who employed him, who wanted a clearer sound, and the ancient monuments authorities, who wanted unimpeded visibility. As he explained, the transparent veil that covered the church at the level of the entablature, and had been developed by him to prevent the sound being lost in the vault, reflected a compromise between their opposed needs. But most revealing was his discussion of the building's susceptibility to reverberations. As originally built and decorated, the structure, like many ancient religious buildings, had a great echo. This had been perfect for priests who wanted to impress their congregations, but the length of the reverberations produced so reduced the clarity of sound as to make the church unsuitable for modern concert performances. His solution, especially the combination of a veil over the vault with a massive triangular baffle suspended over the entrance, reduced the reverberation, especially that of the low notes, to an acceptable level. What he did not say is that in doing so they also diminished the "sublimity" of the structure. Not only had the building lost some of its visual infinity and uniformity, but the music now had an excessive clarity. As Burke said:

> It is one thing to make an idea clear and another to make it *affecting* to the imagination . . . . and so far is a clearness of imagery from being absolutely necessary to an influence upon the passions, that they may be considerably operated upon without presenting any

image at all, by certain sounds adapted to that purpose; of which we have a sufficient proof in the acknowledged and powerful effects of instrumental music. In reality a great clearness helps but little towards affecting the passions, as it is in some sort an enemy of all enthusiasms whatsoever.[18]

The requirements of modern acoustics were in robust conflict with the eighteenth-century notion of the sublime.

The performance thus presented an object lesson in sublimity. The scale of the building, its uniformity, its darkness, and the noisiness and complexity of the performance were all well calculated to create a sublime effect. So, too, was its spoken introduction, with its evocation of astronomic magnitude. But while the science of that introduction was truly sublime, as were the uniformity and infinity suggested by the regular beat of the pulsar's radiation, which was beamed into the church direct from Jodrell Bank Observatory, the whole experience was not. The acoustic engineer had so enhanced clarity and reduced the length of the reverberations, especially of the low notes expressive of great force, that the sublimity of the overall effect was considerably diminished. The properties with which the church had been endowed by Fischer von Erlach half a century before Burke wrote his *Enquiry*, properties that had been well calculated to express God's overwhelming power, were drastically infringed, even for a visitor who was exposed to music that had an astronomic dimension and an enormous volume. Neuroscience helps us to understand why.

Notes

1. Edmund Burke, *A Philosophical Enquiry into the Origin of Our Ideas of the Sublime and the Beautiful*, 2nd ed., 1759 (Aldershot: Scolar Press, 1970), pt. I, §§xiii, p. 71.

2. Burke, *Enquiry*, pt. II, §ii, p. 98.

3. Burke, *Enquiry*, pt. I, §§vi and vii, pp. 58 and 59.

4. Charles Darwin, *The Expression of the Emotions in Man and Animals* (London: Murray, 1872), p. 66.

5. Darwin, *Expression*, p. 80.

6. Immanuel Kant, *Critique of Judgement*, 1st ed., 1793, bk. 2, §29, in James Creed Meredith, *Kant's Critique of Aesthetic Judgement* (Oxford: Oxford University Press, 1911), p. 130.

7. Kant, *Critique*, §23, pp. 92 and 93.

8. Burke, *Enquiry*, pt. II, §v, p. 119.

9. Burke, *Enquiry*, pt. I, §v, p. 57.

10. Kant, *Critique*, §24, p. 94.

11. Semir Zeki, *Inner Vision* (Oxford: Oxford University Press, 1999), pp. 199–202.

12. Burke, *Enquiry*, pt. II, §ix, p. 132.

13. Burke, *Enquiry*, pt. II, §ix, p. 133.

14. Burke, *Enquiry*, pt. II, §iii, p. 99.

15. Burke, *Enquiry*, pt. II, §xvii, p. 151.

16. Kant, *Critique*, §§25–27.

17. Kant, *Critique*, §25, p. 98.

18. Burke, *Enquiry*, pt. II, §iii, pp. 102–103.

7

Quantum Romanticism

The Aesthetics of the Sublime in David Bohm's Philosophy of Physics

IAN GREIG

According to Kant, aesthetic experience is central to mediating the relationship between humans and the world. Through aesthetic judgment, we experience the harmonious working of our rational faculties and then project that harmony outward onto the empirical world, wherein we perceive in objects the formal unity that we discover in ourselves. While this harmony can impress upon us the intelligibility and purposiveness of nature, at other times we are overcome by nature's infinite greatness and renounce the attempt to understand it. Judgments of the sublime, in particular, reveal our inner metaphysical infinitude, sensed as a moral absolute within us or an aesthetic intimation of the same. Whereas beauty is recognized through an accordance of imagination and understanding, which generates a feeling of peace and harmony, recognition of the sublime requires completion by reason, whereupon imagination's insufficiency is overridden by reason's capacity to conceive of totality as a whole, evoking a feeling of unity and oneness that highlights what Kant calls our supersensible or noumenal self. Thus, we do not remain passive in the presence of totality but transcend the province of the senses and appreciate the grandeur of nature as a grandeur of the mind; in Israel Knox's words, "a grandeur born of Reason and the consciousness of moral worth."[1] Ultimately, for Kant, it is this supersensible aspect that is sublime.

In this chapter I suggest that similar sentiments can be discerned within the discourse of contemporary physics. Especially in the popular literature, much of the writing on quantum theory and cosmology reflects an aesthetic appraisal of the sublimity of nature that revives the romanticist link between nature and aesthetics first established by Kant in the *Critique of Judgement*. In particular, the sublime may provide the aesthetic justification for the transcendent impulses that sometimes accompany discussions of one of the central tenets of quantum physics—the idea that "the physical world is one unbroken, undissectable, dynamic whole."[2]

The impact on the imagination of ideas stemming from physics is demonstrated by a burgeoning literature that testifies to the enduring capacity of quantum mechanics to constantly initiate speculation that far exceeds its instrumentalist efficacy. Ever since the realization in the mid-1920s that nature, in Werner Heisenberg's words, could "possibly be as absurd as it seemed to us in these atomic experiments,"[3] quantum mechanics has introduced some of the most far-reaching ideas in contemporary metaphysics. Overturning the classical view of a rational and mechanistic world having an objective existence "out there," independent of the observer, quantum theory challenges assumptions of an intuitively imaginable and knowable order in the universe and continues to raise deep and unresolved questions regarding the nature of reality and our place within it.

Prominent amongst those engaging with such questions is the British theoretical physicist David Bohm, who has developed a distinctive and idiosyncratic philosophy founded on the proposition of an undivided universe. Based on his understanding of quantum mechanics, Bohm's holistic conception of an "implicate order" governing all phenomena goes beyond physics to embrace the whole of life. Although his views are not shared by the majority of physicists, Bohm's ideas are often acknowledged by those seeking in science an alternative worldview that endorses our connectedness with nature as a whole. Characterized by themes of transcendence and order, of the whole and the parts, the infinite and the finite, these discussions disclose a longing for unity with a realm beyond the world of appearances that has historically preoccupied all metaphysical inquiry. Why quantum physics, in particular, should stimulate such nonphysical speculation is a question that may be more readily understood in aesthetic terms as a response to the ineffable that accords with Kant's rationale for the sublime.

As Kant pointed out, our conception of the infinite universe as a unified whole is inextricably tied to the awareness of our own finite existence. In demonstrating the depth and intensity of cognitive experience, the sublime, with its emphasis on the supersensible, elicits the awareness of our potential to rise above the limitations of finite phenomenal existence and to apprehend a transcendent, noumenal realm. While in rational terms this may be seen as the triumph of reason over sensibility, imputations of a transcendent reality beyond sensory reach point to that other field to which both the discourse of the sublime and the discourse of physics have historically turned for authentication: theology. Permeated by an implicit theism governed by the inscription on the Western mind of the transcendental status of infinity—both mathematical and metaphysical—physics and the sublime are each characterized by the ancient desire for knowledge of the One. Indeed, the belief that the study of the laws of physics is an act of communion with God, a sacred union between mind and nature via mathematics, recurs throughout the line of rational thought leading from classical physics to the intangible mathematical entities of today. At the same time, the tradition of the sublime has its origins in the *mysterium tremendum* associated with religious experience. The influence of the Christian mystical tradition can be discerned in the romantics' echo of Pseudo-Dionysius' assertion that the sublime experience reveals God. Described by Knox as a feeling of "spiritual rehabilitation," the exultant inner spark elicited by the sublime renders it first and foremost a tradition of spiritual inquiry—as Jack Voller puts it, "an aesthetically founded quest devoted to recovering intimations of the divine."[4]

Against this backdrop, physics's relationship to the sublime is embedded within a framework defined by romanticism. While in a descriptive sense the sublime may apply only to the supersensible, its aesthetic experience can be evoked by natural phenomena, and for the romantics the sublime manifested through contemplation of the spectacles of nature, where awe and admiration at the overwhelming power and majesty of nature elevated the soul beyond the world of the commonplace. At the same time, the natural sublime established the significance of inner space wherein the immeasurability of physical space was metaphorically linked to the infinitude of our supersensible faculty. By providing a substitute for Christian cosmology displaced by the new sciences, the eighteenth-century sublime countered the anxieties elicited by the apparent withdrawal of God from human affairs by associating the infinite universe with the majesty of the divine, an aesthetic response wherein the boundless universe came to signify the infinite power of God. The experience of the infinite, then, serves as a correlation of transcendence, the spatial or temporal enactment of consciousness in search of that which is beyond itself, the "other."[5]

The regulative metaphors of romanticism's understanding of the world are reappearing through the terms in which physics is perceived today. To contemporary romantics, physics reveals that "the cosmos is a seamless unity—the One—which contains within itself, in potential form . . . all that which is possible or impossible, all that is or might be."[6] Indeed, the encounter with the unpresentable in physics is driving a shift to the poetic margins of the discourse and beyond, whereupon a kind of literary romanticism inflects much nontechnical writing on quantum theory as physicists attempt to come to terms with a new level of description in which everything is interconnected in an unvisualizable whole. While the wave–particle duality defies the imagination, in quantum field theory material substance dissolves into a formless void of ephemeral quantum interactions, a dynamic network of vibrating energy patterns in which all points of the cosmos are connected via the quantum foam. The view that the underlying structure of the universe is field-like rather than particle-like, along with the Copenhagen interpretation's assertion that quantum reality cannot be said to exist independently of observation, has led to the embodiment in physics of metaphysical themes attributed to the feature of "wholeness" of quantum phenomena, and the inseparability of subject and object, which are instituting a dramatic reappraisal of how we describe the universe.

And with each new discovery, the quotient of accessible sublimity in physics is continually being enlarged. While holism in one form or another has always been a tenet of modern physics, the implications of nonlocality stemming from Bell's Theorem and the 1982 experiments of Alain Aspect attest to a whole that ultimately has no definable parts, since science can only correlate relations between particulars.[7] The realization that the whole cannot be represented as a sum of its parts has disclosed an unexpected limit to scientific knowledge that sees ontological and metaphysical speculation interacting with other areas of thought. Consequently, the appeal to terms used in transcendental, theological, and mystical discourses by physicists and philosophers attempting to "explain" quantum mechanics and its epistemological implications to the wider audience not only indicates just how entangled physics and metaphysics are but also confirms John Wheeler's observation that, not content with insights into nature, we are demanding of physics some understanding of existence itself.[8]

However, the separation of theory and experiment that characterizes physics today means that it attempts to speak on behalf of all aspects of human endeavor at a time when it has reached its empirical limits. Insofar as it deals with the "real," the fact that quantum theory cannot account for the ontological status of its constituent entities and, in its orthodox interpretation, can make no assertions about what is ultimately real has resulted in its enlistment in aid of all sorts of metaphysical positions. At the same time, the increasing separation of the objects of physics from the realm of the real is resulting in the crystallization of a distinctive sphere of experience within physics itself—that of the aesthetic.

Specifically, aesthetics manifests in our attempts to make sense of the world, in our efforts to discover or, indeed, *create* meaning as imaginatively participating perceivers. Such a bias is particularly evident in Bohm's outlook. Indeed, Bohm articulates a position that might be termed "quantum aestheticism." Throughout his career, Bohm was preoccupied with understanding the meaning of quantum mechanics. Dissatisfied with the conventional, mechanistic formulations of physics, in which all phenomena are viewed in a strictly external relationship to one another, he challenged prevailing orthodoxy by formulating an alternative theory to the Copenhagen interpretation that seeks to restore what he

David Bohm

felt was the natural symmetry between humans and the world. Although Bohm's "ontological interpretation" of quantum mechanics remains controversial, of interest here is the interplay not only between physics and metaphysics in his views but between physics and aesthetics, manifest in his desire to establish a coherent meaning for the whole. Observing that "physics is more like quantum organism than quantum mechanics,"[9] Bohm calls for a reinstatement of meaning and value into the scientific worldview and extends his ideas on physics into a range of previously nonphysical areas such as perception, language, creativity, and society to propose a unity of all human experience in which consciousness and the creative imagination are intrinsic features of the whole.

Characterized by a combination of science and metaphysics, of explanation and speculation, of the rational and the mystical, Bohm's outlook, with its focus on nature *and* beyond, discloses a romantic longing to embrace the whole of experience while at the same time acknowledging that the universe may remain forever inconceivable and our knowledge always incomplete.

Bohm's position is predicated on nonlocality's inference of the undivided wholeness of the universe but also conditioned by the awareness that this wholeness as a totality is unpresentable. As he puts it, physical theories "are not 'descriptions of reality as it is' but, rather, ever-changing forms of insight, which can point to or indicate a reality that is implicit and not describable or specifiable in its totality." To regard our theories as truth rather than as a way of viewing the world reinforces a fragmentary picture of the world, he says, leading to the illusion that the world is constituted of separately existing "atomic building blocks." Bohm advocates that we adopt a new form of insight in which the world is viewed as an "*undivided whole,* in which all parts of the universe . . . merge and unite into one totality."[10]

In acknowledging the existence of an indivisible whole that may forever lie beyond the reach of observation and measurement, physics enters the realm of the sublime. According to Kant, the sublime is a feeling generated by a confrontation of the mind with an object that defies assimilation by the senses, an object that threatens to overwhelm our perceptual and imaginative capacities. Such is the cosmos. Nevertheless, the aim of physics is to incorporate all parts of the universe, from the elementary particles within atoms to the largest astronomical structures, into a single conceptual framework. Indeed, the goal, in Einstein's words, is nothing less than a theory "whose object is the *totality* of all physical appearances," whereupon "the whole of physics would become a complete system of thought."[11] In attempting to present a totality that we are unable to comprehend as a totality at the level of the senses but that we can comprehend at the level of thought, physics implicitly encodes within its discourse the aesthetic spark of the sublime. That is, by endeavoring to encompass this complex totality within a single conceptual framework, physics provides a mechanism whereby our inability to assimilate this totality at the sensory level is overridden by an awareness that we can assimilate it in rational terms. Such a capacity enables us to feel the scope and superiority of our rational nature. This is the source of our aesthetic pleasure. Or, as physicist Paul Davies puts it: "The ability of physics to unify the strange and bewildering world about us cannot fail to be profoundly inspiring."[12]

Bohm's inspiration is manifest in his "implicate order" hypothesis—a new generative order emerging from discoveries at the quantum level that he claims has significance extending far beyond physics to an understanding of unity in all areas of life. Developed to explain the mysterious phenomenon of particle "entanglement" stemming from the Einstein–Podolsky–Rosen (or EPR) paradox, where two particles issuing from a single source appear to instantaneously "communicate" with each other after being separated by vast distances, Bohm's theory proposes the existence of a deeper realm of unobserved subquantum forces underlying and acting upon the quantum field, a hidden dimension of infinite depth that ultimately gives rise to the material space-time universe. In this view, interactions between subatomic particles, although they manifest in discrete units, are more like dynamic and interlinked structures that are grounded in the whole from which they unfold. It is a conception in which all phenomena are interrelated. Relying on analogy as much as any formal schema for its description, Bohm invokes the idea of a hologram to describe the implicate order as "a vast dimensionality, a much richer sort of reality," wherein each point in spacetime is not only connected to every other but contains every other via a constant process of enfolding and unfolding that he calls the "holomovement." This movement is described by the mathematics of quantum theory, he says, "an unbroken and undivided movement of waves that unfold and enfold throughout the whole of space."

By its very nature, the implicate order is an abstract and elusive concept. It is, in essence, an aestheticized model of reality in which everything is enfolded within an infinite totality—an attempt, in terms of the sublime, to represent the unrepresentable. Reversing the reductionist view that the whole is the sum of its parts, Bohm proposes instead that the whole is fundamental and that the parts, the "explicate order," are the result of the enfolding of the underlying implicate order, the primary framework of reality. For Bohm, the idea that the world is one dynamic and undivided whole has now been established in physics, and as this new order of a whole and undivided totality replaces the old order of separation traditionally utilized by science, it will become the basis for a new worldview—a synthesis of science, art, and spirituality that can enable the establishment of a common and unified culture.

In evoking a hidden synthesis that supposedly reveals the inner connection between things, developments in physics are said by some commentators to be contributing to our rediscovery of the "aesthetic and spiritual meanings of nature" and our connectedness to a "creative and mysterious universe."[13] Indeed, Bohm sees physics as confirming the intuitive sense of unity with nature as a whole that he felt as a child: "When I first studied quantum mechanics I felt again that sense of internal relationship—that it was describing something I was experiencing directly."[14] Characterized by themes of longing and fulfillment, the romanticist belief that a communion with nature can provide direct access to intuitive truth now finds expression in physicists' holistic accounts of an interconnected universe.

This discourse also confirms that an abiding influence in modern metaphysics remains Kant's vision of a finite humanity cast into an infinite world yet exalted by the infinitude of freedom, wherein our conception of the infinite universe as an all-encompassing unified whole is bound to the poignant self-consciousness of the finitude of our own existence. The desire for a sense of unity in the face of the unpresentable accords the subject a central role in the creation of meaning, and Bohm's comments express that same tension between our awareness of finitude and our status as rational beings that characterizes the Kantian sublime. And while Kant declared science incapable of solving the problems of metaphysics, since science deals only with the world of appearances—the world of phenomena and not the world of noumena or things-in-themselves—Bohm is transforming metaphysical problems into problems of physics.

Bohm wants to restore objective reality to the conventional understanding of quantum mechanics. Rejecting the nonrealist inferences of the Copenhagen interpretation, Bohm insists that the wholeness implicit in a reality structure founded on quantum discontinuity, context-dependent form (the wave-particle duality), and nonlocality is actual, that it matters, and that it warrants serious investigation. He contends that the experimental implications of quantum mechanics have been suppressed in favor of a commitment to mechanism, despite being philosophically inconsistent with the experimental facts, and that as the nature of reality dissolved into something indescribable, physicists abandoned any attempt to grasp the world as an intuitively comprehensible whole and instead restricted themselves to developing a workable mathematical formalism. He asserts that because the structure of reality implied by twentieth-century physics is enfolded within the whole, it is thus available to human experience. As such, wholeness is more than a theoretical notion for Bohm—it is a field of meaning, a living totality that includes us as active participants.

Any conception of reality, however, that is expressed in terms of a dynamic, seamless continuum in which the observer is an active participant cannot be entirely understood by reason alone but requires mediation via the aesthetic imagination. As in the romanticist view of the world as an infinite text in which our interpretations derive from our "sympoetic" readings of it, Bohm regards the question of the meaning of a given set of facts and equations in physics as finally an aesthetic one, involving modes of appraisal and perception that are "basically artistic in nature." The key to understanding in science, he says, lies in recognizing the significance of those acts of creative imagination that accord meaning to the interpretation of data: "Physics is a form of insight, and as such it is a form of art."[15]

For Bohm, the aesthetic considerations of science extend to the very fabric of experience. In an echo of Friedrich Wilhelm Schelling's contention that nature and art are the product of the same activity, one that is aesthetic in essence (the only difference between them being that with the world the creative activity is unconscious whereas with art it is conscious), Bohm declares that not only is scientific inquiry itself richly aesthetic, but "fundamentally, all activity is an art . . . art is present everywhere"; even the idea of the cosmos (derived from the Greek "order") is "an artistic concept really." And holding that imaginative insight is a prerequisite in the search for truth, he asserts that the perception of new relationships in quantum mechanics hinges on a process of creative discernment that not only informs scientific methodology but dissolves any distinction between art and science. Consequently, the unity between these "different but complementary" modes of experience reveals a harmony that enables insights into questions of meaning and value that are "at the basis of humanity's process of assimilating *all* experience into one dynamic and creative totality."

That Bohm should embrace physics in order to explore the meaning and value of existence is consistent with the romanticist elevation of nature as a means to self-reflection, wherein the individual is not only the source of reason but also the source of creativity. In opposing Newtonian reductionism and the empiricist concentration on the part at the expense of the whole, the romantics applied the "union of the senses and imagination" to those questions on which science was largely silent: value, aesthetics, and the unity of all things. Influenced by Schelling's metaphysical natural philosophy, which combined Kant's transcendental idealism with the belief that transcendental ideas had an objective existence in nature, the aesthetic contemplation of nature was thought to restore a fragmented humankind to unity, reintegrate sense and feeling with reason, and relocate the individual in an organic relationship with their environment. Kant's link between natural science and aesthetics as quests for order that lie beyond the bounds of sense found acceptance in the belief of Goethe and others that aesthetic appraisal involves a reenactment of the processes of creation through which one may intuit the "inner ground of nature."[16]

In claiming that the holistic implications of quantum physics signal a new and creative mode of human thinking, Bohm is extending themes that have informed the relationship between philosophy and aesthetics in the modern period into the discourse of quantum mechanics. In its ability to satisfy our desire for meaning, Bohm sees science as a symbol of the morally "good" in much the same way the romantics regarded aesthetic appraisal; both modes of experience reveal intimations of the transcendental in the purposiveness and intelligibility of nature. The capacity of science to assimilate what Bohm calls the "structural relationships of existence" enables insights into coherent "essential relationships" that can illuminate nature as a whole. Our appreciation of this coherence, he says, produces feelings of harmony, oneness, and beauty, which parallel the state of the universe as a unified totality.

Bohm's belief in an infinite and harmonious universe imbued with a hidden order belies an implicit aesthetic of the sublime that finds expression in his notion of the holomovement—the infinite and unlimited fundamental ground of all being. Characterized by themes of unity and separation, order and fragmentation, Bohm's description of the holomovement evokes a conception of the sublime in which the field of the finite, tangible to the senses, is suspended within the field of the infinite, beyond space and time and the current conceptual grasp of physics. Enfolding the implicate order within the wave structure of the universe-in-motion, Bohm's holomovement refers to a level of reality concealed below the level of experience, an unknowable and indescribable sea of energy that cannot be defined explicitly but can only be known implicitly. Although not present to perception, this "undivided wholeness in flowing movement" may be sensed like the vastness of space is sensed, says Bohm, as an emptiness or nothingness. The holomovement is a universal flux that is prior to the things that form and dissolve in it and in which we are like that which we observe, merging ultimately with the universal field movement. In this flux, mind and matter are not separate substances. Such a totality collapses any distinction between thought and reality, he says, enigmatically declaring "that *reality is no thing* and . . . also not the *totality of all things.*"

Rather than covering some limited, measurable, and completely knowable domain, the project of physics is, observes Bohm, "an indefinite and unending unfolding into a measureless unknown." Here, issues of "totality" in science intersect with Kant's theme of the limit to reason, where the idea of an absolutely unknowable reality beyond the world of experience—the realm of the noumenal, or "thing-in-itself"—induces the sublime awareness that something transcends the mind's capacity to grasp it. Inferences of ultimate unknowability in science require that all differentiated systems in nature be supplemented, in theory and fact, by other systems. And while each of the systems isolated in the study of nature is in some sense a whole, no single system, except the entire universe, can fully realize the cosmic order of the totality. As such, the unbroken and undivided whole cannot be limited in any specifiable way, says Bohm: "It is not required to conform to any particular order, or to be bounded by any particular measure . . . *the holomovement is undefinable and immeasurable.*"

In occasioning imaginative representations that strive after something that lies beyond the bounds of experience, Bohm's philosophy of science demonstrates that the quest for the absolute possesses universal characteristics. His metaphysical notion of an underlying unbroken wholeness, of "that-which-is," has resonated with theologians who, seeking in science a confirmation of their own view of the world, have interpreted the holomovement as an aspect of the transcendent features of nature that correspond to a divine presence, or as the pure consciousness of God. From a physicist's perspective, Jack Sarfatti combines Bohm's holographic universe with Wheeler's proposition of a participatory universe to propose a union of mind and universe as one immense, cognitional multidimensional projection space in which the wave function of the universe provides the mechanism for communion with God.[17]

Here, at the poetic extremes of quantum speculation, science intersects the spiritual via the aesthetic perception of the infinite in the finite. Appeals to the theological and the ineffable are a particular mark of the confluence of physics and mysticism that characterizes much popular quantum conjecture. Claims that quantum mechanics affirms the creative potential of human consciousness to apprehend a transcendent domain or to glimpse the essential nature of things are, in aesthetic terms, the contemporary expression of the human subject's elevation above the world of the senses. From the enlistment of Eastern spiritual philosophies as heuristic aids in understanding the holistic inferences of physics to a gnostic desire to apprehend and experience the absolute, in quantum mysticism the romantic sublime surfaces in neoplatonic mode, conspiring with notions of *ekstasis* and a unity beyond the realm of the material that reflects a desire not to apprehend the universe as real but, as Plotinus put it, as "one living organism."[18]

Bohm himself was a follower of Krishnamurti, and although he does not explicitly endorse the worldview of quantum mysticism, his writings are a common reference for those outside science who do. However, Bohm's metaphysics comes with a unique and authoritative warrant. Bohm's ontological interpretation of quantum mechanics, also known as the "hidden variables" theory, is considered the main alternative to the Copenhagen interpretation, and some argue that, but for historical contingencies, it might have become dominant among physicists. Seeking to synthesize opposing positions in contemporary physics, the Bohm–Vigier causal theory has been enlisted into the so-called "postquantum" attempt to locate a theory of consciousness within physical reality. With its inferences of a quantum force in the form of a pilot wave, Bohm's theory provides an alternative interpretation of nonlocality that goes beyond the pragmatic applications of conventional quantum mechanics, which might, according to its advocates, enable the unification of mind with matter.[19]

Postquantum mechanics introduces the contemporary paradigm as a product of consciousness. In postquantum holism, there can be no separation between the mind and its object. In this picture, the cosmos emerges as the ultimate feedback loop; the universe itself is conscious, with human consciousness "participating" in this dynamic as an intrinsic feature. The awareness that there is a limit to our ability to fully apprehend physical reality means that acts of cognition can no longer be viewed in classical terms as representations or images of independently existing facts; instead, the physical foundation for the activities of the brain must now be viewed as intimately connected with the whole.

The problem remains, however, that science conditions us into an awareness of the existence of this whole yet cannot affirm its existence in strictly scientific terms.[20] A world that is a dynamic, living whole cannot be represented as a sum of its parts; the reality itself can never be finally disclosed, defined, or described by physical theory, only inferred. Thus, physics has brought us to the threshold of knowledge concerning the existence of the undivided whole, yet we cannot cross that threshold in terms of consciousness. Any experience of reality in which the human is to be regarded as a part of the whole can only be mounted in mystical terms, or what Einstein referred to as "cosmic religious feeling"— or, in other words, the feeling of the sublime. That is, although the extent of this whole revealed by physics cannot be comprehended in terms of sense perceptions or images, it can be conceptualized as an infinite totality wherein reason prevails over that which overwhelms the senses. As Bohm's writings demonstrate, this affirmation of rational comprehension overcomes the possibility of alienation and becomes instead a source of satisfaction, a confirmation of unity with the cosmos through which the limitations of embodiment are felt to be transcended.

Paradoxically, it is our awareness of the limitations of our capacities that provokes our awareness of what lies beyond them. Here, Kant accords aesthetic experience a revelatory status, the implication being that it enables us to view ourselves in relation to a transcendent, noumenal, reality where the grandeur of the world and the inexpressible order that allows us to know it is revealed, along with an awareness of our limitations in the face of the ineffable. Judgments of the sublime, in particular, elicit a harmony between the world and our cognitive and creative capacities that not only forms the basis for the unity of the human subject as a whole but also signals a metaphysical link between the phenomenal and the noumenal, a link that cannot be translated into reasoned argument but can only be *felt*.

However, we should not mistake this transcendent exaltation for the real thing. In light of the delusion, or "subreption," by which the subject mistakes the natural for the ideal, Kant cautioned that the sublime has no objective status—it does not exist externally in nature, but only in the mind. Whereas a mystical experience can be described in terms of *satori*, that instant intuition of the whole as a totality that signals the union of the individual with the Godhead, the limits imposed by our faculties on experience require that judgments of the sublime entail the rational discernment in the face of the infinite that the experience is one of desire for that which is beyond experience—a desire for an absolute "other," expressed as (but not representing) a phenomenologically present but ideal and transcendent reality that can never be attained but only morally or aesthetically intimated.

In pointing to a realm beyond the reach of human powers of knowledge and description, the sublime is epistemologically transcendent in that it denotes the object of experience while leaving open the question of its existence. The aesthetic significance of the experience, then, is determined in terms of the mind's *relation* to the object as one of unattainability, the presentation of which signifies an absence that has historically been taken as a metaphor for "God," "the soul," or "the Absolute," and so forth. In other words, the sublime concerns not the "real" as such, but only the limitations of our attempts to grasp it. Hence, for Kant's "object of nature" we may substitute any object (a physical theory, for example), if the failure to assimilate the object as a whole determines the mind to regard this inability as a sign of the mind's relation to a transcendent order of being.[21]

Physics too, insofar as it deals with the real, seems to have constructed its own ideal realm, a conceptual realm beyond experimental verification where understanding is achieved only via arcane mathematical procedures. Upholding the dualistic conception of reality as abstract disembodied ideas in a domain separate from and superior to that of material objects, physicists often disclose a Platonic mathematical idealism in their talk of touching something universal and eternal in a realm beyond the senses. As such, the Western metaphysical dream of a divine and timeless reason permeates both the discourse of physics and the aesthetic of the sublime; the identification of each with a transcendental presence, an ultimate legislating principle originating outside space and time, has proven irresistible.[22]

In an era skeptical of the claims of essentialist aesthetics, the transcendent aspect of the sublime is its main problematic. However, the endemic theologizing and mysticism in physics, while equally problematic, suggests that the response to the unpresentable possesses universal characteristics. Here, the sublime's enthusiasm for the absolute as the theme for any given representation renders it the appropriate aesthetic for quantum metaphysics. As Bohm's writings attest, judgments of the sublime highlight the active role of the imagination, enabling a differentiation of the sublimity of our moral being from the aesthetic experience of it while at the same time reinforcing our sense of sublimity because we detect its traces in the aesthetic judgment of formless nature. In articulating the phenomenology of our response to a mode of being that can only be mounted in aesthetic terms, the rational credentials of the Kantian sublime afford an ontological nuance to the transcendent impulses accompanying physics's dematerialization of the physical world. The truth is no longer "out there" but, as Bohm demonstrates, the interiorization of the infinite relocates the locus of transcendence to the quantum vacuum within, thus confirming Wheeler's observation: "Never in all our exploration of nature have we come upon a domain more mysterious and more inspiring of thought than the quantum. Nothing sounds stranger, nothing is more revelatory."[23]

The penetration of matter by physics has revealed the universe to be not a collection of objects but, as Davies puts it, "an inseparable web of complicated energy patterns in which no one component has reality independently of the entirety; and included in the entirety is the observer."[24] In its inexorable progress from the visible world into the invisible and on into an increasingly theoretical realm illuminated by the intellect alone, quantum physics is catalyzing a contemporary sublime that finds its expression in the metaphysics of physicists such as Bohm. As the philosopher Paul Crowther has written, the sublime springs from "finite being's struggle to launch itself into and articulate the world."[25] Bohm's implicate order reflects this same "primal urge" into transcendence, that desire to create or discover meaning in the face of the unpresentable that is made vivid in the experience of the sublime.

Notes

1. Israel Knox, *The Aesthetic Theories of Kant, Hegel and Schopenhauer* (New York: Humanities Press, 1958), p. 58.

2. E. E. Harris, "Contemporary Physics and Dialectical Holism," in Richard F. Kitchener (ed.), *The World View of Contemporary Physics* (Albany: State University of New York Press, 1988), p. 158.

3. Werner Heisenberg, *Physics and Philosophy* (New York: Harper, 1958), p. 42.

4. J. Voller, "Neuromanticism: Cyberspace and the Sublime," *Extrapolations* 34, no. 1 (1993), p. 18.

5. Voller, "Neuromanticism," p. 19. See also E. Tuveson, "Space, Deity, and the Natural Sublime," *Modern Language Quarterly* 12, no. 1 (1951), p. 22.

6. Roger S. Jones, *Physics as Metaphor* (Minneapolis: University of Minnesota Press, 1982), p. 4.

7. In 1964, John Bell confirmed mathematically that nonlocality was an inherent feature of the quantum mechanical description of nature. Bell's Theorem has since been upheld by a series of experiments, beginning with Aspect, Dalibard, and Roger's 1982 Paris experiments. See www.fdavidpeat.com/bibliography/essays/healtech.htm (19 October 2010).

8. John A. Wheeler and Wojciech H. Zurek, *Quantum Theory and Measurement* (Princeton: Princeton University Press, 1983), p. 210.

9. F. David Peat and John Briggs, Interview with David Bohm, *Omni* (January 1987), p. 5. www.fdavidpeat.com/interviews/bohm.htm (7 July 2009).

10. Unless otherwise indicated, all quotes of Bohm are taken from his book *Wholeness and the Implicate Order* (London: Routledge, 1980). See also Bohm and F. David Peat, *Science, Order, and Creativity* (London: Routledge, 1987); Bohm and B. J. Hiley, *The Undivided Universe* (London: Routledge, 1993); and F. David Peat, *Infinite Potential: The Life and Times of David Bohm* (Reading, MA.: Addison-Wesley, 1997). For an edited collection of Bohm's writings on the implicate order see Lee Nichol, *The Essential David Bohm* (London: Routledge, 2003); and for an overview of Bohm's aesthetics see Lee Nichol's compilation, *On Creativity* (London: Routledge, 1998).

11. Albert Einstein, cited in John D. Barrow, *Theories of Everything: The Quest for Ultimate Explanation* (Oxford Clarendon Press, 1991), p. 89.

12. Paul Davies, *Superforce* (London: Unwin Hyman, 1985), p. 1.

13. Charles Jencks, *The Architecture of the Jumping Universe; A Polemic: How Complexity Science Is Changing Architecture and Culture* (London: Academy Editions, 1995), p. 23. Also quoted in Richard Coyne, *Technoromanticism: Digital Narrative, Holism and the Romance of the Real* (Cambridge, MA: MIT Press, 1999), p. 105.

14. F. David Peat and John Briggs, Interview with David Bohm, *Omni* (January 1987), p. 3, www.fdavidpeat.com/interviews/bohm/htm (July 7 2009).

15. Bohm (1979), quoted in Paul Davies, *God and the New Physics* (London: Dent, 1983), p. 129.

16. Andrew Cunningham and Nicholas Jardine (eds), *Romanticism and the Sciences* (Cambridge: Cambridge University Press, 1990), p. 90.

17. See Michael Talbot, *Mysticism and the New Physics*, (New York: Bantam, 1980), pp. 60–62; and Jack Sarfatti, *Matter, Mind and God*, www.raven1.net/mcf/hambone/nmg.html (12 October 2010).

18. Plotinus, quoted in G. A. Turnbull, (ed.), *The Essence of Plotinus*, (1948), p. 80, cited in Coyne, *Technoromanticism*, p. 55.

19. See, for instance, Henry Stapp, *Mind, Matter and Quantum Mechanics* (Berlin: Springer, 1993).

20. See Minas Kafatosand Robert Nadeau, *The Conscious Universe: Part and Whole in Modern Physical Theory* (New York: Springer, 1990).

21. See Guy Sircello, "How Is a Theory of the Sublime Possible?" *Journal of Aesthetics and Art Criticism* (Fall 1993), pp. 542–550; Louis Wirth Marvick, *Mallarmé and the Sublime* (Albany: State University of New York Press, 1986), pp. 3, 157; Thomas Weiskel, *The Romantic Sublime: Studies in the Structure and Psychology of Transcendence* (Baltimore, MD: John Hopkins University Press, 1976), p. 23.

22. See Brian Rotman, *Ad Infinitum: The Ghost in Turing's Machine* (Stanford, CA: Stanford University Press, 1993), p. 57.

23. J. A. Wheeler, "Law Without Law," in Peter Medawar and Julius Shelley (eds.) *Structure in Science and Art* (Amsterdam: Excerpta Medica, 1980), p. 145.

24. Davies, *Superforce*, p. 49.

25. Paul Crowther, *The Kantian Sublime: From Morality to Art* (Oxford: Clarendon Press, 1989), p. 167.

8

Disobedient Machines

Animation and Autonomy

SCOTT BUKATMAN

Cinematic narrative has, from its inception, been intrigued by tales of the unnatural creation of life, an unnatural creation that often allegorized the cinematic apparatus itself. In tone, however, films like *Pinocchio* (1940), *Gertie the Dinosaur* (1914), and the *Out of the Inkwell* cartoons are quite different from other tales of synthesized life, such as *Frankenstein* (1931), *Metropolis* (1926), and more recent science fictions, including *2001: A Space Odyssey* (1968) and *Blade Runner* (1982). All these works tell of the creation of a being or beings that, at some moment, begin to act autonomously; Pinocchio, Gertie, Out of the Inkwell's Koko the Clown, Frankenstein's monster, Maria, Hal, and the various replicants all become what might be called "disobedient machines." But while some creatures, Gertie and Pinocchio among them, are allowed their trespasses, for other disobedient machines the consequences can be dire. (In this essay I concentrate on relatively early instances of the creation of synthetic life in the cinema, works that are, not coincidentally, most devoted to detailing the act of creation itself.)

Perhaps these tales can be usefully distinguished from one another by their alignment with two rhetorical modes: the uncanny and the sublime. Some of these disobedient machines are uncontainable, *sublimely* terrifying rather than *uncannily* disturbing. The sublime and the uncanny are closely related: both stage a confrontation with the limits of human power and agency, and both are heavily freighted with the weight of the unknown. And yet they operate at different scales, raise different questions, play through different conflicts, and align with different aesthetics. The sublime figures the unknown as excess; the uncanny re-presents the familiar in terms of estrangement. Both inaugurate crises for the subject: the sublime appears and is resolved as an epistemological crisis around the limits of human knowledge; the uncanny instantiates an existential crisis centered upon the unknowable interiority of the self. The uncanny is preoccupied with undecidability, the porous boundaries between human and nonhuman, organic and inorganic.

Anthony Vidler has pointed to the close relation between the two modes: "Aesthetically an outgrowth of the Burkean sublime," the uncanny is "a domesticated version of absolute terror, to be experienced in the comfort of the home and relegated to the minor genre of the *Märchen* or fairy tale."[1] In the uncanny, argues Tzvetan Todorov, "events are related which may be readily accounted for by the laws of reason, but which are, in one way or another, incredible, extraordinary, shocking, singular, disturbing or unexpected, and which thereby provoke in the character and in the reader a reaction similar to that which works of the fantastic have made familiar."[2] The uncanny is rooted in the conundrums of logic and rationality, while the very nature of the sublime pushes beyond those bounds.

The creation of artificial life, in both literature and film, is often fraught with elements of the sublime. Such stories nearly always engage the discourse of *things man was not meant to know*; not for nothing is *Frankenstein* subtitled *The Modern Prometheus*. Synthetic creation narratives are about, among other things, the harnessing of power—perhaps an inappropriate power—a power that dazzles and seduces. In these stories, as elsewhere, the sublime stages a confrontation between man and the limits of his power; the human is diminished (or struck dumb) by an awareness of a power greater, by far, than his own. Burke argued that the sublime is constituted through the combined sensations of astonishment, terror, and awe. It initiates a crisis in the subject by threatening human thought, habitual signifying systems, and, finally, human prowess. However, in Kant's formulation, man was also exalted by the encounter with the sublime, through his ability to cognize the existence of so vast a power. To imagine something greater than oneself does, after all, call upon remarkable powers of reason and imagination. The nineteenth and twentieth centuries added the newly unleashed forces of technology to produce what Leo Marx has famously labeled "the rhetoric of the technological sublime,"[3] a rhetoric that was aligned in America with the sense of manifest destiny. Despite being the creation of mankind, technology presented a majestic and terrifying power that exceeded that of any individual—not just in its output, but in its very conception and execution. (Of course, these being human products, after all, the human is exalted still further—but the terrifying prospect of technology run amok is inevitably present.)

In many films, the creation of synthetic life becomes an occasion for sublime excess. The arcing light and bubbling beakers involved in the creation of the robot Maria recurs in the storm-tossed skies and the crackling of lightning that presage the animation of Frankenstein's monster and his ineffable bride in James Whale's two Frankenstein films (immeasurably aided by Franz Waxman's florid orchestrations). And there is King Kong: the synthetic creation of animator Willis O'Brien. The narrative presents Kong as *discovered* in nature rather than *created* in a laboratory, and so he might seem misplaced here, among the diegetic synthetic figures of Maria and Frankenstein's monster, and yet the film is obsessively centered upon its own spectacle of synthetic life. (Does the film's advertising slogan, "The Eighth Wonder of the World," refer to the giant ape—a throwback to prehistoric wonders—or to the miniature model that is brought to life through the modern miracle of cinema?) The sublime excess of the synthetic life narrative pervades *Kong*. The shrouded mysteries of Skull Island speak for themselves, and after his capture he is, just like Maria before him, "premiered" before a cosmopolitan crowd. The staccato popping of blinding flashbulbs in this sequence replaces the atmospheric spectacle and the montage of scientific contraptions found in the other films, and, just as the first evidence of life in *Frankenstein* was the monster's twitching hand, so it is Kong's arm that first breaks free of the onstage constraints that bind him. The spectacle of the inanimate creature "coming to life" cuts across these films, accompanied by as much *Sturm und Drang* as the filmmakers can muster.

Mad scientists are sublime, or they at least venture into the territory of the sublime. There is alchemy at work, or sorcery: some kind of transgression against the laws of matter and man. The mad scientist operates on that border between a properly scientific knowledge and prohibited, taboo theories and methods. The films fairly crackle with demonic energies, energies that also seem to be the lifeblood of the creators themselves. Henry Frankenstein frantically scrambles about his cavernous laboratory, a manic gleam in his eye, in a sequence that culminates with his hysterical shriek of "It's alive!" Carl Denham pontificates at breakneck pace throughout the first half of *King Kong*, verbally flattening everything in his path. And the mad inventor Rottwang in *Metropolis*, already partly artificial, masterfully manipulates the switches and levers like a symphony conductor. Indeed, there is another sense in which we can see these figures as "conductors," in that their own demonic energies seem to be transferred to their very creations. Denham and Rottwang virtually disappear from the scene after the birth of their creations, and Henry lies catatonic. In every case, the animatedness of the creator seems to infect, or be conducted to, his creation.

Compared with these demagogues and demigods, an early newspaper cartoonist and animator like Winsor McCay cuts a benign figure, affable and sociable. Yet in some ways his *Little Nemo* film, which features one of the earliest examples of animated drawings, can be seen as an exemplary cinematic narrative involving the creation of life.[4] The film, after all, begins with McCay making the ridiculous claim that he can make his pictures, his characters, move, a claim to which his friends respond with warm mockery. Most of the film's brief running time is devoted to chronicling the process of this creation, which allegorizes the cinematic apparatus by recovering its essence as a series of still images imbued with movement through the act of projection. (The resemblance is even more pronounced in *Gertie the Dinosaur*, with its "reanimation" of a prehistoric monster, albeit a rather playful one.) Like Frankenstein and Rottwang, McCay wields the power to bring the inanimate (the drawing) or the prehistoric (the dinosaur) to life. Here, too, the creator's own ferocious energy seems to transfer to his drawings—after the Nemo animation begins, McCay disappears from the scene. The drawings not only take on properties of life, but seem to obviate the necessity of a creator at all.

But it is difficult to imagine spectators reacting to the animation of the trickster Flip in *Little Nemo* or the adorable Gertie with the same terror inspired by Frankenstein's monster or the robot Maria. Nor does one hear about McCay's films producing the sort of shock that presumably greeted the film *OF A TRAIN*

ARRIVING at La Ciotat AS RECORDED BY THE Lumière Cinématographe. Instead, it seems likely that the effect was of something playful, marvelous—remarkable, surely, but hardly the stuff of a paradigm shift. They belong more firmly to the realm of the *uncanny* than the *sublime*. If artificial life narratives frequently reference the rhetoric of the sublime through their invocation of the dark realms beyond everyday reality, they also, almost inevitably, summon a sense of the uncanny. The doubled figure of creator/creation, the shadow figure that haunts the original, the familiar returned in the guise of the unfamiliar: these are common tropes of uncanny representation. After all, as a character remarks in Karel Capek's 1920 robot saga *RUR*, "nothing is stranger to man than his own image."

Automata are especially closely bound to the uncanny. Wilhelm Jentsch posited that a feeling of uncanniness could arise in an encounter with, among other things, clockwork automata: "In telling a story one of the most successful devices for easily creating uncanny effects is to leave the reader in uncertainty whether a particular figure in the story is a human being or an automaton."[5] And, as the sublime has the uncanny as its more intimate double, so does the technological sublime have a counterpart in a technological uncanny. The double belongs to the field of the uncanny, but the automaton belongs to the technological uncanny—specifically, a *mechanical* uncanny. While automata might deceive the eye or ear, their motive force is neither mysterious nor alchemical, but rather mechanical. They are engineered creations, often produced by a solitary figure who is something more of a tinkerer than a man obsessed: Jacques de Vaucanson, *Pinocchio*'s Geppetto, Spallanzani, the inventor in ETA Hoffmann's tale of "The Sandman," the Edison of *The Future Eve*. These are not mad scientists; they do not tap demonic or sublime energies.

The sublime as a phenomenon is aligned with the gigantic, while the uncanny is more easily aligned with the miniature. Kong, Maria, and Frankenstein's monster all bear the mark of excess and gigantism in their very beings as well as their sites of origin: the gargantuan Kong and the tribal rites of his Skull Island home; Maria's mesmeric powers, which she shares with the vast and glittering Metropolis itself; the Monster's horrific visage, and the mountaintop laboratory where he is given life. As Susan Stewart has written, "the gigantic represents infinity, exteriority, the public, and the overly natural . . . the gigantic unleashes a vast and 'natural' creativity that bears within it the capacity for (self)-destruction."[6] The monster is gigantic, while the automaton is something more of a miniature. The miniature is characterized by "clockwork precision," Stewart notes;[7] it "represents a mental world of proportion, control, and balance," while the gigantic "presents a physical world of disorder and disproportion."[8] The miniature further represents "closure, interiority, the domestic, and the overly cultural."[9] The sublime is expansive, large-scaled, and cosmic; the uncanny is an altogether quieter and more intimate mode. (Sigmund Freud's own example, picked up by Gaston Bachelard, was the childhood home.) Its operative mode is less slack-jawed wonder than a subtle *frisson*. The sublime is best appreciated from the safety of distance, while the uncanny is an altogether more proximate and intimate phenomenon.

Fundamental to Jentsch's description of the uncanny is a state of undecidability. The uncanny, Vidler writes is, "in its aesthetic dimension, a representation of a mental state of projection that precisely elides the boundaries of the real and the unreal in order to provoke a disturbing ambiguity, a slippage between waking and dreaming."[10] Or between material and immaterial, organic and inorganic, human and almost-human. The romantic figure of the double—the shadow self, the reflection, the man in the crowd, the monster who will "be with you on your wedding night"—haunts the ego by threatening its boundaries, or reminds us of that time when our ego boundaries were not yet fixed. The automaton represents still further levels of undecidability: Does it have a soul? Can man create life in his own image, in imitation of God? What "is" the human? Here is the ultimate instance of the *unheimlich*: we are no longer *at home* in our own bodies.

The cinema, with its phantasmatic doublings of the real world and the people in it, is a fundament of the technological uncanny, as Laura Mulvey has argued. Early cinema produced its sense of the uncanny in varied ways that could be aligned with the ideas of both Freud and Jentsch. Mulvey writes, "If the contemporary response to the Lumière films aligns them on side of Freud's ghostly uncanny, Méliès transfers to cinema many characteristic attributes of Jentsch's uncanny, exploiting technological novelty as well as the cinema's ability to blur the boundary between the animate and the inanimate with trick photography."[11]

The literary historian Michele Bloom argues that the cinema was the locus for a reinvigoration of a Pygmalionesque desire that transcended any particular narrative iteration. "The very medium," she writes, "embodies the long-standing human desire for the animation of the inanimate. Even when the Pygmalion paradigm fails in film, the medium itself succeeds in creating the illusion of movement."[12] The cinema quickly discovered the mysteries of stop-motion trick films and films that animated a series of drawings, and animation has persisted (and arguably has become the dominant mode) in the era of digital production, although it remains associated most strongly with fantasy and children's entertainment.

If cinema presented itself as life's uncanny double, it could be argued that animation was the fantastic, refracted double of the photographed profilmic, the uncanny's uncanny. As Lev Manovich notes,

> Once the cinema was stabilized as a technology, it cut all references to its origins in artifice. Everything that characterized moving pictures before the twentieth century—the manual construction of images, loop actions, the discrete nature of space and movement—was delegated to cinema's bastard relative, its supplement and shadow—animation. Twentieth-century animation became a depository for nineteenth-century moving-image techniques left behind by the cinema.[13]

For Mulvey, however, live-action cinema epitomized the uncanny properties of the medium precisely because it moved animation *away* from the realm of drawings: "From the perspective of the uncanny, the arrival of celluloid moving pictures constitutes a decisive moment. It was only then that the reality of photography fused with mechanical movement, hitherto restricted to animated pictures, to reproduce the illusion of life itself that is essential to the cinema."[14] In this view, the cinema is fundamentally connected to the "undecidability" of the moving figure—most fully in its most realist incarnations. Perhaps the fundamental uncanniness of the cinema is repressed via animation by being safely relegated to, and contained within, the realms of childlike fantasy and cartoonish irreality—analogous to those "minor genres" to which Vidler saw the uncanny confined.

Bloom traces the history of what she calls the "Pygmalionesque imagination" and its translation from the failed Pygmalion tales of nineteenth-century literature (e.g., Hoffmann's "The Sandman") to its twentieth-century rebirth in the cinema, where some of the earliest films were of artist's models coming miraculously (and harmlessly) to life. Gaby Wood finds a similar transposition in the trick films of Georges Méliès: "In Méliès's workshop, you might say, automata gave birth to the movies."[15] Much of this original novelty disappeared or went "underground" as the cinema became more familiar, but perhaps one place where medium's originary uncanniness continued to lurk was in the realm of the animated film.

Animation holds a privileged position in the history of film in that it subtends and predates the moment of cinema's emergence. Animated drawings were the basis of such prephotographic media as the thaumatropes, phenakistoscopes, zoetropes, and the like. Automata and phenakistoscopes alike embodied issues of *anima*: objects were mechanically charged with movement and seeming life. Sergei Eisenstein's writing on animation, in particular the early cartoons of the Walt Disney studio, emphasizes the liberating sense of *anima*. His notebooks record his interest in the word "anima," with its implication that all objects possess a natural life or vital force. In early animation, everything pulsates with life—not only in the foreground, where barnyard animals, steamboats, and airplanes bounce and stretch in zesty rhythm, but, in the days before cel animation, even the "stable" background forms seemed to possess a vibrant, vibrating buzz of their own. For Eisenstein, this animism spoke to an inherent dynamism of form: everything was in the process of becoming something else; the world was mobile, in all senses of the word. His discussion of animism is also strongly evocative of the complexities of the uncanny:

> The degree to which—not in a logically conscious aspect, but in a sensuously perceiving one—we too are subject every minute to this very same phenomenon, becomes evident from our perception of the "living" drawings of none other than Disney.
> We *know* that they are . . . drawings, and not living beings.
> We *know* that they are . . . projections of drawings on a screen.

> We *know* that they are . . . "miracles" and tricks of technology, that such beings don't really exist.
> But at the same time:
> We *sense* them as alive.
> We *sense* them as moving, as active.
> We *sense* them as existing and even thinking![16]

There are few better examples of the pervasive animism to which Eisenstein is so devoted than the early films of Winsor McCay, especially perhaps in his first film, in which he endowed his *Little Nemo* characters with movement and, consequently, life.

Early animation is filled with creations that disobey their creators. The first of McCay's characters to come to life in his *Little Nemo* film is Flip, the mischief-maker, and in his next film Gertie (the dinosaur) has moments of disobedience—McCay even calls her a "bad girl." In Max Fleischer's animated series, the uncontainable Koko the Clown emerges out of the inkwell at the start of each cartoon, an inkwell to which he often has to be forcibly returned by the cartoon's producer. Why is this? If the animator can create a figure who can do his precise bidding, and make him jump through as many hoops as he pleases, why create something that immediately rebels?

The rambunctious creations of early animation not only *disobey* their creators; in many cases, they actually *replace* them. We watch McCay produce drawing after drawing in his *Nemo* film, but once the animation begins he is not seen again. The *Out of the Inkwell* cartoons often begin with Max Fleischer in his studio, but once Koko appears, Max, too, often fades from the scene. The labor that goes into producing the drawings—a labor that we initially were privileged to watch—is erased once that "spark of life" energizes the animated figure; a transfer of energy occurs: the creation is energized, and the creator enervated.[17] As the history of animated film progressed, the animator, whose presence informed the films of Émile Cohl, McCay, the Fleischer brothers, and others, recedes from view. The onscreen production of images, a production that fully established the authority of the artist, is superseded by increasingly autonomous characters.

In his indispensable study of the early animation industry, Donald Crafton emphasizes this diegetic shift. "The 'hand of the artist' disappears," he writes, "its place now occupied by characters who become agents of his will and ideas and through which his presence is known. They are his amanuensis."[18] But, despite their becoming "agents of his will," this transition in emphasis from animator to character also serves to repress the animator(s) as producer. The cartoon character comes to provide the continuity and centering function once furnished by the figure of the animator. The images and the life they possess begin to seem self-generated, springing spontaneously into existence with no evident progenitor. Taylorist production is superseded by images and beings that seem to generate spontaneously.

But to borrow Tom Gunning's language, perhaps the animator does not entirely disappear but rather "goes underground." Here two works from the Disney studio might illustrate the continuing concern with the illusions of animation, the desire of animators, and the autonomy of disobedient beings. In *Pinocchio*, the studio perhaps achieved its realist masterpiece: a film rich in visual and kinetic details, with a greater unity and consistency than its first feature, *Snow White and the Seven Dwarves* (1937). The film featured enormously detailed backgrounds, sophisticated lighting and atmospheric effects, and superb character animation but still retained some heterogeneity: cartoony animation for most characters, but rotoscoping for the Blue Fairy (a detail to which I return).

While it is something of a commonplace to criticize the Disney studio's sanitized adaptations of classic children's literature, the best of the films, and *Pinocchio* is surely one of those, can be fascinating in their own right. What makes *Pinocchio* so fascinating is not simply its hallucinatory realism (is that an oxymoron?), but its profound reflexivity. This tale of a little wooden boy becomes a metaphor for the aspirations of the Disney studio itself, and a longer standing dream of animation in the sense discussed above. This is entirely in keeping with Eisenstein's sense that Disney represented all that animation could do. What is Pinocchio *about*? Geppetto is a master craftsman who wants to accomplish something more: he wants his creation to be real. What was the task of the Disney animators? To create animated characters that were able to stimulate audiences' identification and empathy, and *belief*. Geppetto's workshop is filled with clockwork automata that erupt in a riot of limited but cacophonous action (See Fig. 11). The automata, in their repetitive, mechanical actions, are reminiscent of the simple looping image sequences of an earlier era's optical toys, while Pinocchio himself, with his greater range of movement, motivation, and expressivity, is more properly cinematic, from the moment the Blue Fairy animates him, well before he becomes a "real" boy.

The Pinocchio who stars in the book by Carlo Collodi is rather different from Disney's version. (He is, not to put too fine a point on it, kind of a prick—after the Talking Cricket warns him to be a good and obedient boy, Pinocchio throws a wooden hammer at him and kills him.) The most significant difference between the book and the Disney version is that literature's Pinocchio is already sentient even before he is carved into the form of a puppet, when he is nothing more than a log. The film's Pinocchio begins as an inert marionette who is animated only by the combination of Geppetto's desire and the Blue Fairy's intervention (just as the cinematic Pinocchio begins "life" as a series of inert drawings, to be granted life by the cinema itself). It is significant that the means of realizing Geppetto's desire within the diegesis is the aforementioned Blue Fairy, who is herself rotoscoped—she's a hyperreal hybrid of live action and animation who embodies an uncanny blurring of boundaries (See Fig. 12).

What is it that makes Pinocchio *a real boy*? The easiest answer, for both the book and the film, is *morality*—morality defines the human. The ostensible moral is that disobedient boys will come to no good in the world. The human is not defined by the ability to *reason* (lying, after all, involves pretty good reasoning) or *emote* (he can emote right away), but by the ability to make (or want to make) correct *moral* choices. What keeps him from being a real boy, by this reading, is therefore his disobedience. But I propose a different reading, one that would put Pinocchio into a longer history of "disobedient machines." I want to propose that it is precisely Pinocchio's *mis*behavior that makes him real, that makes him human.

Here we return to the determinedly disobedient creatures of early animation—why must they disobey? Perhaps it is precisely because something that simply follows its programmed instructions will never be anything *more* than an automaton. The spark of life that separates automaton from living being is precisely its assertion of autonomy. The creation that disobeys does not just come to life; it takes on a life *of its own*. That is, the machine behaves autonomously and *proves* its autonomy by misbehaving. Rebellion is an assertion of self, an assertion of free will in more or less playful terms. As Pinocchio sings in defiance of Geppetto and school and Jiminy Cricket and responsibility in general, "There are no strings on me."

In the introduction to her edited volume *Genesis Redux*, Jessica Riskin discusses Michelangelo's representation of the creation of life on the ceiling of the Sistine Chapel, with Adam's outstretched arm nearly touching the hand of God: "There, between the two fingers, one aiming, the other waiting, is Michelangelo's representation of life itself, reaching from God to Adam. It is, quite simply, a gap."[19] The "understated drama" of this gap articulates, for Riskin, something of the mysteriousness of the life force, but one could also say that this gap, this separation, between creator and creation is fundamental on another level. Adam himself is, after all, a little like Pinocchio, who had to remove himself from Geppetto, who had to cut the strings that bound him to his creator, to become an autonomous being. Here is Adam, a figure created in the image of God, his creator, who is seemingly predestined to disobey. Original sin, the fall from grace, is the fall *into* humanity. This fallibility, this disobedience might be what makes man less than God, but in other contexts, such as *Pinocchio*, or *Frankenstein*, or even *My Fair Lady*, it makes creations into something more than mere machines.

We can therefore find a three-part division in *Pinocchio*: Geppetto, the divine creator; the slavish automata that populate Geppetto's workshop; and Pinocchio, the disobedient individual. Intriguingly, one can find this identical structure in another Disney saga from the same period, *Fantasia*'s adaptation of *The Sorcerer's Apprentice*, also from 1940, featuring Disney's biggest star, Mickey Mouse. That the animated creature is something uncanny, irrational at its core, and perhaps even a bit magical lies at the heart of "The Sorcerer's Apprentice" section of Disney's *Fantasia*. While most cartoons play out the creation/creator duality on the most playful levels, "The Sorcerer's Apprentice" operates on the darkest levels. Mickey Mouse, once the exemplar of the unruly, animistic spirit so celebrated by Eisenstein, and a character voiced, at first, by Walt himself, plays the hapless apprentice, in over his head. Putting on his master's hat, Mickey commands a broom to fetch water, but then is unable to stop his enthusiastic servant. Splintering the broom with an axe only multiplies the problem, as each piece becomes a new broom, equally dedicated to its task. Only the return of the sorcerer returns the world to its proper order—again, the omnipotent creator, able to bestow the spark of life; the automata, marching in lockstep, forever completing a task that is already complete; and the misbehaving creation, played by that one-time avatar of animistic freedom, Mickey Mouse. Here he is an apprentice, engaged in being schooled, and even the magisterial sorcerer is finally forgiving of his apprentice's trespass, with a hint of an indulgent smile and a swat on the *tuchus*.

"The Sorcerer's Apprentice" belongs squarely to both the uncanny *and* the sublime. (The brooms belong, at least at first, to the former, the magisterial wizard and the growing sense of terror to the latter.) It is worth noting that the brooms, despite their compelling anthropomorphism, function only as mindless drones, quite the obverse of the attentive and independent Gertie or Koko. It is also interesting that multiplicity, the very foundation of frame animation's operations, is revealed (the one broom, itself composed of a series of drawings, breaks apart into a series of brooms), only to become an object of fear. It also stands as something of a conservative tract regarding the Disney studio's own mission statement: animism is powerful stuff and should remain under the control of seasoned professionals. How far we have come from *Steamboat Willie*.

Conclusion

And so tales of synthetic life veer from the playful to the nightmarish.[20] What begins to distinguish these two registers? A central issue for these protagonist-creators and the beings they have generated is *control*: Does the creation elude the creator's control, and if so, in what way? And what are the consequences of this slippage for the creator and his creation? The synthetic being can demonstrate its autonomy by behaving unpredictably: most obviously through overt disobedience, perhaps running amok across the landscape (*Frankenstein*) or refusing orders (*Pinocchio*). The creator might be at risk at the hands of his creation (*Frankenstein*) or might suffer indirectly for the creation's behavior (*Pinocchio* again).

Automata, on the other hand, do not rebel: their very name precludes the rebelliousness that signifies the beginnings of autonomous being. If they are mistaken for human, this is usually the result of human error. (The creators themselves rarely make that error, of course, but may encourage others to do so, as in the case of Hoffmann's Spalanzani.) Automata *enchant* through their resemblance to the human: Olympia beguiles the gullible Nathaniel in "The Sandman," just as the animated drawings of Gertie have beguiled audiences for nearly as long as cinema has existed. Thus, the distinction between the darker, more sublime tales of synthetic life and the more playful iterations of uncanniness can be fairly well aligned with the contrast between *misbehavior* and *misperception*.

Outside the realm of fiction, automata do not quite fool us— they only threaten to. The first time one beholds the breathing Sleeping Beauty at Mme. Tussaud's in London, one looks, then looks again, surprised by the barely perceptible motion of the figure's chest, but the illusion is quickly detected, and after testing one's assumption (perhaps by surreptitiously poking it), one can feel amused and then move on. Similarly, flat animated characters like Gertie and Pinocchio, marvelous as they are, do not resemble the human too much. Despite the uncanny pleasure derived from viewing an automaton with some of the properties of a living being, the viewer's sense of real and unreal was never in question. On the other hand, the narratives of *Metropolis, Frankenstein, King Kong*, and more recent films such as *Blade Runner* all present automata (robots, monsters, and replicants) sharing the diegetic space with humans. All these creations are brought into a world populated by humans; they exist within a simulacrum of our reality. They are the kinds of beings that Mary Douglas and Noel Carroll have referred to as "categorical mistakes" that literally cross over from, transgress, the field of representation, and take up a place in the realm of being.[21]

One emblematic shot stands out: during Kong's New York rampage, he is seen through a window, looking in at a sleeping woman who just might be the one he seeks. Kong is framed in the window like something from a Freudian dream-screen. In the next moment, however, Kong's giant paw enters from the side of the frame (presumably through another window), seizing the poor brunette and dragging her off. What this quite literal transgression of the screen space demonstrates is that these figures are not just disobedient in their behavior—they are disobedient in the very fact of their being, which precludes their being trained, constrained, or contained. They must, instead, be destroyed, leaving the borders restored.[22]

That these are categorical mistakes fits nicely with Mulvey's sense that live-action cinema was a new instantiation of the uncanny. While moving drawings could be safely relegated to the realm of toys and handicraft, too unreal to matter, the cinema represented something both more *automatic* and more *phantasmatic*. Cinema demanded of its audience a continual (and almost willful) misperception on the most fundamental level: illusions of movement, depth, and presentness.

BEYOND THE FINITE

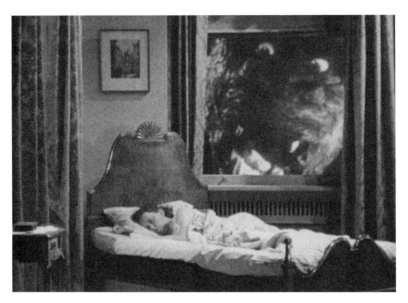

King Kong at the window, from *King Kong* (1933)
© *RKO Radio Pictures*

Those synthetic life narratives which introduce fantastic characters into the midst of a world of humans might then be seen to stage *a recuperation of that misperception*. These stories stage the possibility of the misperception turning out to be correct. *I know very well but even so* . . . is the formula for fetishistic satisfaction (and it should go without saying that automata serve as exemplary fetish objects). By subverting the safe remove of *"I know very well"* in favor of that haunting *"but even so,"* these fictions give us an excess of presence, a plenitude that is, seemingly by definition, monstrous. This might begin to account for the slippage between the uncanny and the sublime in narratives of synthetically created life: while the cinema may be innately uncanny, tales of synthetic life frequently borrow the trappings of the sublime. The undecidability of the uncanny is resituated within a narrative context that asserts the reality of the illusion, replacing uncertainty with a more sublime and transgressive terror.

Notes

1. Vidler, Anthony, *The Architectural Uncanny: Essays in the Modern Unhomely* (Cambridge, MA: MIT Press, 1992), p. 3.

2. Todorov, Tzvetan, *The Fantastic: A Structural Approach to a Literary Genre*, trans. Richard Howard (Ithaca, NY: Cornell University Press, 1975), p. 46.

3. Marx, Leo, *The Machine in the Garden: Technology and the Pastoral Ideal in America* (New York: Oxford University Press, 1964), p. 195.

4. The animated sequence was originally prepared to accompany his vaudeville act, and the live-action framing sequence, supervised by J. Stuart Blackton, was added later for theatrical release.

5. Cited in Sigmund Freud, "The 'Uncanny'," in *Writings on Art and Literature* (Palo Alto, CA: Stanford University Press, 1997), p. 202.

6. Stewart, Susan, *On Longing: Narratives of the Miniature, the Gigantic, the Souvenir, the Collection* (Durham, NC: Duke University Press, 1993), p. 73.

7. Stewart, *On Longing*, p. 94.

8. Stewart, *On Longing*, p. 74.

9. Stewart, *On Longing*, p. 70.

10. Vidler, *The Architectural Uncanny*, p. 11.

11. Mulvey, Laura, *Death 24X a Second: Stillness and the Moving Image* (London: Reaktion Books, 2006), p. 46.

12. Bloom, Michelle E., "Pygmalionesque Delusions and Illusions of Movement: Animation From Hoffmann to Truffaut," *Comparative Literature* 52, no. 4 (2000): 292.

13. Manovich, Lev, *The Language of New Media* (Cambridge, MA: MIT Press, 2001), p. 298.

14. Mulvey, *Death 24X a Second*, p. 52.

15. Wood, Gaby, *Edison's Eve: A Magical History of the Quest for Mechanical Life* (New York: Anchor Books, 2003), p. 183.

16. Leyda, Jay (ed.), *Eisenstein on Disney* (Calcutta: Seagull Books, 1986), pp. 54–55. Eisenstein here distinguishes between two kinds of knowledge, even giving sensual knowledge a kind of priority. But he also describes a not unpleasurable confusion in the viewing subject. Animation represents a "categorical mistake" in the sense that Mary Douglas described it in *Purity and Danger: An Analysis of the Concepts of Pollution and Taboo* (London: Routledge, 1984), what we know to be inanimate takes on a simulacrum of life. More on the categorical mistake at the end of this essay.

17. Of course, there was a pragmatic reason for showing the creation of the cartoon in early animation: without a demonstration of how it worked, audiences would not understand what they were seeing.

18. Crafton, Donald. *Before Mickey: The Animated Film 1898–1928* (Chicago: University of Chicago Press, 1993), p. 298.

19. Riskin, Jessica, "Introduction: The Sistine Gap," in *Genesis Redux: Essays in the History and Philosophy of Artificial Life*, ed. Jessica Riskin (Chicago: University of Chicago, 2007), p. 1.

20. At first glance, Disney's *Pinocchio* would seem to belong to the realm of the playful and the uncanny, rather than the sublime and terrifying—Pinocchio is no monster. But when one considers the dark consequences of the little wooden boy's behavior, it becomes arguably the most traumatic tale of all.

21. Douglas, *Purity and Danger*; Carroll, Noël, *The Philosophy of Horror, Or Paradoxes of the Heart* (New York: Routledge, 1990).

22. There is an admixture of live-action and animation at the end of *Gertie the Dinosaur*, but in this instance it is the real person (McCay) who crosses over *into* the animated realm—this clearly does not pose the same threat as Kong's unboundaried rampage.

9

On the Sublime in Science

ROALD HOFFMANN

No Place for the Sublime?

When the tillers in the field of science look at what artists and philosophers have thought of the sublime, I would hazard the guess that they find little of seeming relevance. The rhetorical strategies for approaching the sublime, dear to Longinus, would seem to have no intrinsic value to a system for acquiring reliable knowledge that seeks for infinitely paraphrasable universals. Not that scientists eschew rhetoric (today's excesses expressed as "hype," a word with a tie to the Greek *hypsous* for the sublime). Edmund Burke's evocation of the avoidance of pain, of "a sort of delightful horror, a sort of tranquility tinged with terror,"[1] at the root of the sublime, is too poetic for that part of science that flattens emotion. And while disinterested scrutiny and contemplation, a *sine qua non* of philosophical aesthetic theory, are most definitely part of science, what if that intent observation of the wonders of the world is but to lead one to God, as in Psalm 19?

> The Heavens declare the glory of God,
> The sky proclaims his handiwork.

Or, as Ralph Waldo Emerson says, "The noblest ministry of nature is to stand as the apparition of God"[2]?

The religious clearly scares scientists. So the molecule or theory that in another time might have been said by the person who made it to be given by the grace of God, now comes courtesy of serendipity. Of course, people have always found ways to imply, and not so subtly, either, that there were reasons God or chance favored *them*. For scientists, Pasteur's paraphrase of the serendipity creed, "Chance only favors the prepared mind,"[3] has served very well in this respect. The practices of science lead away from expressions of awe. Our stock-in-trade, the scientific article, that ossified, neutered, and ritualized way the new is communicated, certainly does not allow it.

So where is wonder voiced? And how? For the most part, scientists, feigning humility, would prefer popularizers to do it for them. When moved with a carrot or a stick to speak to the general public, scientists do dare to express their encounter with the sublime. They put on the mantle of the creator of wonders easily—who wouldn't? But often (like poor poets) they do not particularize the claim.

And chemistry, my central science, well—it has a special problem with awe. We do not have the very small of elementary particles or the soaring large of galaxies. Chemistry lacks that easy ladder to the sublime of boundlessness, of the downward or outward freeways to infinity. But then, it has its wonders (as any human activity does) and, as I hope to show, quite interesting ways of contemplating them.

Scientists are continually confronted with astounding novelty, emerging from experiment or theoretical contemplation. Even the words for describing these new things may be lacking. Why not, then, embrace the wonder of feeling that grips one on looking at the workings of the ribosome or the atmosphere, or the beauty of a molecule shaped like a Ferris wheel, or a distant dust cloud across a galaxy? Why not accept that point when one feels one knows something that is really deep and universal? Or has made a molecule—part tinkered, part designed—almost always beyond what one thought one could make. And crafted that understanding, that molecule, in a way that brings the scientist in harmony with others and the universe? Why not come to peace with the sublime?

I will try, in what follows.

Something, Sublime

Words need to be felt. And chemistry is eminently practical. So . . . we need to see something sublime. To feel it, Imagine a beaker with iodine (at bottom) being heated, a dish with ice at the top. (See Fig. 13) This is the middle of an experiment—at the beginning, the solid dark purple crystals of iodine (you usually know it in solution) were all at the bottom of the beaker. How shall I show what happened—with cartoon conventions, a blur, a sequence of still images, a movie, a flip book? Your imagination is fine; the crystals at the bottom of the flask diminish; they simultaneously begin to grow on the cold surface at the top. They grow out of the vapor of iodine, a light purple color filling the beaker.

But this is just a trick—a play on the ambiguity of the word "sublime"! Yet think about it: ambiguity is the stuff of poetry, not of science. And the etymology of the Latinate *sublime*, "under the lintel," is identical and the usage contemporaneous for both the philosophical *and* the chemical sense of the word. That vapor condensed up there. If not above the lintel, certainly at the top of the beaker.

Sometimes we are so enmeshed in the mechanics of science that we shelve our sense of wonder—here at a solid going to a gas to a solid again (even more mysterious looking when the gas is not colored as it is for iodine), without passing through a liquid. The process is described well in the trade by a "phase diagram" and is associated with ideas of the first great American theoretical physicist and chemist, J. Willard Gibbs. But it's no less wondrous.

It's good, then, to have an outsider to our profession remind us of the simple magic of stinks, bangs, and colors, as Oliver Sacks recently did in *Uncle Tungsten*.[4] Let us look at some functional aspects of my science, chemistry, and see where they touch upon the sublime. Along the way we will find that some classical philosophical ideas about the sublime may need revision when confronted with the realities of scientists dealing with what is beyond ken. A good place to begin is with what brought things up high.

Raise the Energies of the Soul

Chemistry was and is the art, craft, and business of substances and their transformations. And now that we have learned to look inside the innards of the beast, there has emerged a parallel microscopic perspective—chemistry is the art, craft, business, and science of persistent groupings of atoms called molecules. The sublime aspect of that microscopic perspective is that it was attained without waiting for any microscopes to show us what is in there. Our certainty grew out of a piecewise, rickety, fuzzy knowing without seeing, with attendant lessons for the philosophy of science.

Mixed up as these macro- and microscopic views are in the nicely illogical mind of the chemist, change, the defining essence of chemistry, is central. And chemical reactions, often seemingly magical, yet part of everyday experience since ancient times, are most assuredly exemplars of essential change. A boiled egg is not an unboiled egg. Such apparent examples of change were what hermetic philosophies of transformation needed to enter into people's minds. So alchemy took up chemistry as a symbol, and then as a practice, and was co-opted by it; alchemists became pretty good chemists.

Then and now, energies were necessary to effect desired change. Even when something should happen, be it love or chemistry, barriers need be overcome. In chemistry (before I get into trouble) it may be light or electricity that activates. But then, and now, Heraclitus had it right—fire is the prime motive force. What actually happens is pretty much what Kant wrote of sublime objects, that they "raise the energies of the soul above their accustomed height."

But I am actually talking about another principle of the sublime, the exercisable *potential* for change. How easy is true change for matter! How difficult it is for us! Which is why people will always be ambivalent about chemistry.

Let me expand on this. Energy may be stored in chemical bonds, to be released in a chemical reaction. To set things into motion, to effect change, one has to overcome some barrier, often a great one. From high school chemistry to the Hindenburg disaster, we have seen explosive mixtures of hydrogen and air just sit there . . . until a match is brought near. Then change takes place (with a vengeance), fundamental change—water, the product of that explosive reaction, is different from hydrogen and oxygen.

But Kant wrote of a soul, high, yet moored in the psyche, not of exploding balloons or turquoise crystals precipitating in a flask. In our mind, change is viewed more ambiguously than it is in the laboratory—do we want to change our job, our partner, our home? There may be good psychobiological and evolutionary reasons for our ambiguity about change—that we desire it yet are afraid of it. Chemistry is most definitely about change. So I would argue that, subconsciously, chemistry, symbolic of change, will always be viewed ambiguously, because of the resonances it sets up in our psyche with other kinds of transformations.

In a science, chemistry, where the emphasis has subtly shifted with time from analysis to synthesis, from discovery to creation (not without attendant ethical problems), we have crafted a sublime dynamic of change, by fire. Chemists understand the arrows in Paul Klee's art. (See Fig. 14)

The Median Sublime

Recognition of the centrality of change, and of the potential for storage of the energies released in reaction, leads to a different perspective on the seemingly quiescent middle. For some, the middle may be the den of compromise and indecision. Or the entropy-driven stillness at the end of things. But the living middle (of human beings and molecules in equilibrium) is tense, for in it is the potential for change, the perturbation of an equilibrium by human action.

People and the press want extremes; I praise the middle and find the sublime in it. In the middle is chemistry, with its tense molecules—tense because they can hurt or heal, sometimes even one and the same molecule: take ozone (a shield from the ultraviolet high in the atmosphere, a component of photochemical smog at sea level) or morphine (the analgesic of choice, as anyone who has had an operation knows, and addictive).

There is more that gives life to the middle. Every molecule is suspended in a multidimensional space defined by crisscrossing polarities: the just-mentioned harm or benefit is one, natural and unnatural another, and what is to be revealed or concealed about a newly made molecule yet another. Matter may be pure

On the Sublime in Science

or completely mixed up; a molecule may be the same or not the same as another molecule. The extreme positions on these axes are interesting only momentarily, I would say, in a kind of "world record" mode. But the tension of being in the middle, for instance—of being impure, yes, but really capable of being more or less pure—that potentiality is the source of motion, of life.

Giovanni Pico della Mirandola, in his "Oration on the Dignity of Man," writes of God:

> Taking man, therefore, this creature of indeterminate image, He set him in the middle of the world and thus spoke to him:
> We have given you, Oh Adam, no visage proper to yourself, nor any endowment properly your own, in order that whatever place, whatever form, whatever gifts you may, with premeditation, select, these same you may have and possess through your own judgment and decision.[5]

I think there is a relationship between the energy stored in a highly strained organic molecule, its C–C–C angles far from the 109.5° of an ideal tetrahedron, and a classical Greek statue, with a pose of one leg bent, one stretched, the upper torso slightly twisted toward us, its prominent musculature. Molecule and statue are both tense; the sublime in our perception of them is in their contained tension.

Natural/Unnatural, on the Road to Diversity

At the root of all is the natural—mind you, not always in the simplistic way we think. So, of the seeds of the plants that through ingenious labor eventually lead to cotton (perhaps the

155

most chemically intensive agricultural crop on earth), rayon (chemically modified cellulose from wood pulp), and nylon (the product of petroleum-based industrial chemistry), the only seed that has a legitimate claim to being truly untouched by human hands is the seed that eventually led to nylon.

There is fecund diversity in the natural—the six components of a moth pheromone, the 900 molecules in the aroma of cabernet sauvignon. And I think this is more than just "beautiful." It is the tangled bank of which Charles Darwin wrote in the last page of *On the Origin of Species*:

> It is interesting to contemplate a tangled bank, clothed with many plants of many kinds, with birds singing on the bushes, with various insects flitting about, and with worms crawling through the damp earth, and to reflect that these elaborately constructed forms, so different from each other, and dependent upon each other in so complex a manner, have all been produced by laws acting around us.[6]

That diversity, amply revealed by chemistry and biology operating in the discovery mode, reifies Kant's idea of the mathematical sublime, the notion of infinite chemical variability. Given time and workings of chance, this incredible variety of molecules that accompanies life evolves—all the greater our grief at species extinction, which could be viewed from the molecular viewpoint as burning a library without reading it.

But chemistry is as much about creation, the synthesis of molecules, as it is about discovery. And so as soon as we see that it takes four "bases" and a five-membered ring molecule, ribose, to which they are attached, to generate a code that instructs for the making of biopolymers, proteins, composed from 20 or so amino acids, we wonder—why four bases, why a five-membered ring, why 20 amino acids? And since chemical transformation is easy, we embark on the synthetic, man- or woman-made, artifactual, unnatural exploration of the (endless) changes that can be wrought upon the building blocks of these biopolymers. The ingenuity in exploring the natural and unnatural universes accessible to us is sublime.

Useful

Because transformation of the natural is so central, and ultimately easy, it leads people impelled by profit and curiosity to create a multitude of unnatural molecules. I suspect Kant would have approved of the unnaturalness of it all; at the same time, he would have disqualified from aspiration to the sublime the ringing of the changes that a pharmaceutical chemist does on an antimalarial drug (to reduce toxicity, improve effectiveness)—because of the purposiveness of it all. Kant might have been wrong here, failing to see the way human beings subvert all categories on their way to the new.

Utility has great transformative power. The selfish and the altruistic are both at work in any human activity, and especially in acts of scientific and artistic creation. People are driven to make the good and the beautiful—an antitumor agent that fits just right into a receptor on a tumor cell—because of its meliorative purpose, and because of potential profit.

In a masterful essay, M. H. Abrams notes that "for some two thousand years of theoretical concern with these matters, it occurred to no thinker to claim that a human artifact is to be contemplated disinterestedly, for its own sake, as its own end for its internal values, without reference to things, human beings, purposes, or effects outside its sufficient and autonomous self."[7]

Abrams traces the English and German routes to the detached contemplative model of aesthetic judgment, and concludes that "when we turn to *King Lear*, or Michelangelo's *Pietà*, or Beethoven's *Ninth Symphony*, or Picasso's *Guernica*, the concepts of art-as-such are patently inadequate to account for the range of our responses to these works, which implicate our knowledge and convictions about the world, our moral interests, and our deepest human concerns."[8]

Making It

I have hinted at the special role of synthesis in chemistry. Now let me make it explicit, for I think this is chemistry's high road to the sublime. The molecular science has moved from analysis, the finding out of what we have—be it a sample of moon rock, the aroma of fresh cocoa, a narcotic off the street—to synthesis. This is the part by design, part by chance making of molecules.

Open a chemistry journal, and what you see left and right is molecules being made. Here it is a laboratory-made natural product, such as vitamin B12; there a stick figure of a cube out of DNA (See Fig. 15)—yes, a synthetic molecule made from pieces of DNA, made because it is a proof of principle that such games can be played with DNA. Transgressing the natural/unnatural divide is an old chemical tradition. Or it could be a molecule constructed for a purpose: to fit a nicotine receptor and block it, or to make a possible superconductor—*Zweckmäßigkeit mit Zweck, pace* Kant. Twenty million new compounds made to date, and we are losing track...

A minor part of the chemical literature actually tests theories, which is what scientists are supposed to do according to the noted philosopher of science, Karl Popper.[9] Most of us create a universe and study the objects of our own creation. As a consequence, the philosophy of science would look very different if it were constructed by practitioners whose training was in modern chemistry.[10]

Philosophers have stressed contemplation or feeling of the sublime, emerging through contemplation of the universe, its forces, mountains, earthquakes, seas. But I think all along there has been another road to the transcendent. This is through the work of creation, the labor of hands and minds combined. It was not done by God in one day. It may be the work of an old house restored, a child raised, a pot glazed sky blue (yes, emerging from fire), or . . . our métier, the making of molecules. Through acts of small human creation, not as often accompanied by ethical judgments as they should be, we carve out the sublime, and so join in the mandate of Genesis.

Albert Eschenmoser, a very wise chemist, reminded me of life. He writes:

> Chemistry definitely has its own and very specific channel to the sublime of boundlessness: the potential of the matter of chemists to have become alive and, once their matter had passed through that barrier, to have evolved and still to evolve into the boundlessness of the living. What are quarks and what are galaxies for if not for being recognized by life? Where would it be, the sublime in the thoughts of philosophers, artists and priests, without that sublime potential of matter, our matter, the chemist's matter?[11]

The reason that I stress synthesis is that it creates a link for chemistry with art.

Beyond Good Science

Much of what we do in science is routine—some is worse, a little is very good. But some soars and is seen as sublime, even in that culture that eschews excesses of emotional praise. What are the characteristics of sublime science? How do we recognize it?

The sublime may be embodied in a paper that is unique, that solves a problem of long standing. This was the case with Andrew Wiles's proof of the proposition that Pierre de Fermat said was true but that he could not fit into the margin of his book. It was true of the Watson and Crick double helical, hydrogen-bonded model of DNA, a structure that immediately explained much and contained immense utilitarian promise.

But the sublime in science may also be diffused and build upon itself. I think this is true of Einstein's four 1905 papers, and of the series of syntheses of natural products by R. B. Woodward and his coworkers over the period 1943–1968, during which they synthesized quinine, cholesterol, cortisone, strychnine, reserpine, chlorophyll, cephalosporin, and colchicine—and eventually (in an equal partnership with Eschenmoser), vitamin B12. Each synthesis was perceived by chemists as a masterpiece for its economy and the novel ways in which chemical bonds were built up. Each synthesis faced seemingly insurmountable obstacles; each found supremely clever ways around them, ways that could be used by others. The syntheses told great stories; Woodward's riveting, beautifully orchestrated five-hour seminars on the syntheses augmented his high writing style. But it is the totality of the crafting of these natural products that seemed sublime.

In Woodward's syntheses a style was created, a way of doing that might have had antecedents, yet was new. It could be emulated, and was. I would argue that this is what the Einstein papers of 1905, in their very different way, did as well. In the Woodward and Einstein papers a *Zeitgeist* (of organic chemistry, of theoretical physics) was formed. Its inchoate pieces were in the air—the statistical thermodynamic arguments that Einstein used so effectively, the beginnings of quantum mechanics that he built on; the organic syntheses of others than Woodward.

Zeitgeists are ephemeral, eminently replaceable. And science is the cult of the new. Their lesson is broadly assimilated, but those original papers of Einstein are hardly read today. The classics of science do not serve as the classics of literature or music; my guess is that less than one in a hundred graduate students in physics has read the Einstein papers. But when one reads them in sequence, not just one, when one goes through the artistic logic of Woodward's natural product syntheses, the reaction of most young people is visceral.

Connectivity, Solace

One feature of this powerful social system of gaining reliable knowledge, modern science, is that its practitioners are compelled to publish, and to do so in abundance. Our stock-in-trade is the scientific article, not the book. In forty-five years I have inflicted on the world more than 500 of these articles; there are people who have published many more.

A typical research chemistry department will also have five to ten guest lectures a week; there are innumerable conferences to attend. Behind all this is an infrastructure of economic support; it's not for nothing that of the $400 million that comes into Cornell for research, $380 million goes into science and engineering ($19.5 million into social sciences, $500,000 for the humanities—peanuts). The desire, necessity of, and addiction to such a flood of communication are made possible by national choices, rational or not. But even in the absence of support, I think it would be there, that urge to tell others what one has done.

Scientific communication represents to me a humanization of the sublime. Let me explain. In a critical mode, I could find much fault in the vast majority of the 500,000 chemical articles published each year. Too many are routine, or worse. But there is a time to praise. Most of those pieces of work reach out. Many are cited by others. Indeed, there is no greater joy than to see your work of use to others, especially those you do not know. Is it because praise is not allowed in the joyless, repressed scientific article, that we seek satisfaction in the citation?

The sublime is shaped in the mind of a human being; it is a solitary construction. Yet it places one in intimate contact with the universe. A neglected part of the sublime is the solace that stepping outside of oneself provides. So Rainer Maria Rilke could write on the flyleaf of the "Duino Elegies" copy he sent to Marina Tsvetaeva in 1926:

> We touch each other. How? With wings that beat
> With very distance touch each other's ken.
> One poet only lives, and now and then
> Who bore him, and who bears him now, will meet.[12]

In another context, it was given to me forty-five years ago, in Nikita Krushchev's time, to study for a year in that Stalinesque wedding cake of a building of Moscow University. I remember to this day the expectation and care, almost reverence, with which my fellow chemistry students took up that issue of the *Journal of Chemical Physics* I brought in. It was, for them, in Soviet Russia, a window on the world, a connection with every scientist out there.

Stepping outside oneself, as solitary as I must be, (I hesitate, for I don't want to deal out the transcendence of paired love, or that of the crowd of Carnival) seems in some way cold. But the sublime has a reconnecting side, that link with other human beings implicit in Rilke's dedication, or in what Marion says at the end of Wim Wenders's *Wings of Desire*, of being alone together. Caspar David Friedrich's paintings often have in them more than one observer.

This is one of the things that science does well. It has mastered the ethic of collaboration. And science's massive communication system connects people. In that tie, there is solace. Unconscious or not, all that read writing (even as Jacques Derrida called it, in another context, "the message that abandons") overcomes isolation and gives meaning to life. The sublime, out of us, seemingly beyond us, is reclaimed by us. Now as a human bond.

Notes

1. Edmund Burke, *A Philosophical Enquiry into the Origin of Our Ideas of the Sublime and Beautiful* (1757), in *On Taste, on the Sublime and Beautiful, On the French Revolution* (New York: Collier, 1909), p. 114.

2. Ralph Waldo Emerson, *Nature* (New York: Thomas Y. Crowell, 1887), p. 55.

3. Louis Pasteur, "Dans les champs de l'observation le hasard ne favorise que les esprits préparés." Lecture presented at the University of Lille, Douai, France, December 7, 1854. H. Petersen, ed, *A Treasury of the World's Great Speeches*, (New York: Simon and Schuster, 1954), p. 473.

4. Oliver W. Sacks, *Uncle Tungsten: Memories of a Chemical Boyhood* (New York: Alfred A. Knopf, 2001).

5. Giovanni Pico della Mirandola, *Oration on the Dignity of Man*, trans. A. Robert Caponigri (Washington, DC: Regnery, 1956), pp. 6–7.

6. Charles Darwin, *On the Origin of Species* (New York: Collier, 1909), ch. 14., p. 528.

7. M. H. Abrams, "Kant and the Theology of Art," *Notre Dame English Journal* 13 (1981): 75–106; p. 76.

8. Ibid., 101.

9. K. R. Popper, *Conjectures and Refutations: The Growth of Scientific Knowledge*, 5th Ed. (New York: Harper & Row, 1974).

10. Roald Hoffmann, "What Might Philosophy of Science Look Like If Chemists Built It?" *Synthese*, 155 (2007): 321–336.

11. Albert Eschenmoser, letter to the author, June 17, 2005.

12. Boris Pastenak, Marina Tsvetaeva, and Rainer Maria Rilke, *Letters, Summer 1926*, trans. Magaret Wettlin and Walter Arndt (San Diego: Harcourt Brace Jovanovich, 1985), p. 27.

Index

Abrams, M. H., 157–58
abstraction, 13
Adam. See Newman, Barnett
aesthetics, 9, 58, 61, 81, 86, 111, 112
  of terror, 15
affective communication instinct, 33
affects, 21–22, 41
  and artistic passions, 34–36
  and cognitive richness, 22
  emotional systems, primal of, 30–31
  human souls and animal souls, 38–41
  and mental life, 22–27
  mind, passions of, 27–29
  mind medicines, search for, 29–30
  soul, neurobiology of, 36–38
  as source of music and language, 32–34
against sublime, 75
  history, of sublime, 78–84
  poor sublime, 86–89
  religious concept and sublime, 84–86
  religious concept in art criticism and history, 86
  science, sublime in, 75–77
  visual art, sublime in, 77–78
agency, 45
akinetic mutism, 32
ancestral mind, 40
ancient sublime, 47
animal souls, 38–41
animation and autonomy, 128–47
  character, 139
  and Disney, 140
  Donald Craft on, 139
  early years, 138, 141
  of Frankenstein's monster, 130
  in history of films, 137
  of inanimate, 135
  Nemo animation, 132
  prehistoric monster, "reanimation" of, 132
  rambunctious creations of, 138
  Sergei Eisenstein's writing on, 137
  of trickster Flip, 132

Index

"Art and Objecthood." *See* Fried, Michael
artistic endeavors
   social emotions and, 33
astronomers, 57, 58
   on color images, 61
   and pretty pictures, 61, 63
Astronomical Journal, 62
automata, 133, 137, 140, 142, 144, 145, 146

Beagle, HMS, 92
beautiful
   and the sublime, 4–5
Beauty and the Contemporary Sublime. *See* Gilbert-Rolfe, Jeremy
Beckley, Bill, 77
   Sticky Sublime, The, 77
Beholding, act of, 48
Bell's Theorem, 110
Bierstadt, Albert, 7
binary model, 14
Blade Runner, 128, 145
Blair, Hugh, 16
Blake, William, 48
Bloom, Michele, 135, 136
Blumenberg, Hans, 44
Bohm, David, 107–8, 111, 112, 113, 114–22, 125
   holomovement, 119
Bohm–Vigier causal theory, 121
Bond, Howard, 64, 70
Boris Godunov. *See* Pushkin, Aleksandr
brain localization, 46
brain maturation, 28
brain molecules, 29–30
brain systems, for maternal devotion, 27

buprenorphine, 30
Burke, Edmund, 4, 12, 15–17, 47, 68, 79, 81, 91–92, 94–97, 101–4, 129, 130, 149
   Philosophical Enquiry into the Origin of Our Ideas of the Sublime and Beautiful, A, 4, 81, 104

Capek, Karel, 133
Cartesian dualism, 37, 39
Chalk Cliffs on Rügen. *See* Friedrich, Caspar David
change, 152–54
characteristic of sublime, 50
chemistry, 150, 151, 152–54, 156
Collodi, Carlo, 140
color images and prettiness, 60, 61, 63
computational theory of mind, 26
concrescence, 54
consciousness, 22, 23, 25, 38, 39, 40, 45, 47, 49, 50, 51, 53, 107, 109, 113, 120, 121, 122
Copenhagen interpretation, 110, 111, 116, 121
core self, 36–37
cortical columns, 26
corticotrophin-releasing factor, 28
cosmic religious feeling, 122
Crafton, Donald, 139
creativity, human, 21
Critique of Judgement, 6, 81, 86, 94, 107. *See also* Kant, Immanuel
Crowther, Paul, 3, 80, 125
   and Kantian sublime, 8

Damasio, Antonio, 24, 51, 53
Dante, 47

Index

Darwin, Charles, 92, 156
   On the Origin of Species, 156
   Expression of the Emotions in Man and Animals, The, 92
Davies, Paul, 114, 125
Death and transience, thoughts of, 7
De Bolla, Peter, 80–81, 82, 84
   Discourse of the Sublime, The, 80
Definition of sublime, 3
De Man, Paul, 83, 84
Denham, Carl, 131, 132
desires, 30
discourse
   on the sublime, 80
   of the sublime, 80
Discourse of the Sublime, The. *See* De Bolla, Peter
Disfiguring: Art, Architecture, Religion. *See* Taylor, Mark C.
Disney, works by
   Fantasia, 142
   Mickey Mouse, 143
   Pinocchio, 128, 139, 140, 142, 144
   Snow White and the Seven Dwarves, 139
   Sorcerer's Apprentice, The, 142, 143
disobedient machines, 128–47
Donougho, Martin, 84
dopamine, 30, 31, 40
"dorsal stream," role of, 98, 99
dual-aspect monism, 23, 24
Ducrot, Oswald, 83
dynamical sublime, 9, 10

Echo Objects: The Cognitive Work of Images. *See* Stafford, Barbara Maria
Edgerton, Samuel Y., 60

Einstein–Podolsky–Rosen (EPR) paradox, 114
Eisenstein, Sergei, 137–38, 140, 142
Elkins, James, 75
   Six Stories from the End of Representation, 75
Emerson, Ralph Waldo, 149
emotional affects, 23
emotional feelings, 21–25, 27, 37–38, 40
emotional systems, 27–28, 37, 39
   most primal of, 30–31
   and seeking system, 31
empathy, 39
endogenous opioids, 29
Enquiry. *See* Burke, Edmund
epinephrine, 40
Erlach, Fischer von, 101, 104
Eschenmoser, Albert, 159, 160
ethical insights stimulation, 16
Expression of the Emotions in Man and Animals, The. *See* Darwin, Charles

Fantasia. *See* Disney, works by
fear, 17
feature detectors, 96
feelings
   affective, 23, 24 (*see also* affects)
   and language, 33
Fermat, Pierre de, 160
Fichte, Johann Gottlieb, 51
first-person consciousness, 50
Fleischer, Max, 138, 139
   Koko the Clown, 128, 138
   Out of the Inkwell, 128, 138
Frankenstein (Shelley), 128, 129, 131, 142, 144, 145
Frattare, Lisa, 64–65

Fried, Michael, 86
  "Art and Objecthood," 86
Friedrich, Caspar David, 7, 15, 80, 163
  Chalk Cliffs on Rügen, 7
functional magnetic resonance imaging (fMRI), 93
Fuseli, Henry, 48, 80

gaseous nebulae
  Hubble images, 62
Gell, Alfred, 45
genes, ancestral voices of, 27–29
Genesis Redux, 141
Gertie the Dinosaur. See McCay, Winsor
Gibbs, J. Willard, 152
Gilbert-Rolfe, Jeremy, 77
  Beauty and the Contemporary Sublime, 77
God, 96, 108, 120, 155
Gödelian swirl of self, 51
Grisey, Gérard, 101
"Große Halle." See Speer, Albert
Guerlac, Suzanne, 83

Harries, Karsten, 14
Heidegger, Martin
  exegesis of Friedrich Hölderlin's poems, 9–10
Heisenberg, Werner, 107
hidden variables theory, 121
history, of sublime, 78–84
Hofstadter, Douglas, 51, 52
Hölderlin, Friedrich, 9–10
Holocaust, 11, 14, 17
holomovement, 115, 119, 120

Hubble images, 57, 62, 64, 65, 69, 70, 71
  aesthetics of, 69
  analysis of the, 57–58
  gaseous nebulae in, 62
  longevity and popularity of, 71
  prettiness, 58, 67
Hubble Space Telescope, 63–64, 67, 68
human children
  desires for physical play, 28–29
human creativity, 21
humanist sublime, 84
human psychiatry, 39–40
human reason, 17–18
human souls, 38–41
Hume, David, 9
  Treatise of Human Nature, 9
Hussein, Saddam, 17

Idea of the Holy, The. See Otto, Rudolf
identity, 49
implicate order hypothesis, 114
information technology, 52
inlaying, 55
Inner Vision. See Zeki, Semir
Iraq war, buildup of, 17
irresponsible sublime, 87

Jentsch, Wilhelm, 133
Jodrell Bank Observatory, 104
Journal of Chemical Physics, 162
joy, 6

Kandinsky, Wassily, 13
  Über das Geistige in der Kunst, 13

Kant, Immanuel, 6, 68, 72, 77, 81, 85, 86, 91, 95, 101, 106, 107, 116
  Critique of Judgement, 6, 81, 86, 94, 107
  influence of Lisbon earthquake (1755), 15
  Observations on the Feeling of the Beautiful and the Sublime, 94
Kantian sublime, 3–4, 7, 10, 78
  of beautiful and the sublime, 5
  of beauty and art, 4
  and Crowther, 8
  on dramatic text, 6–7
  dynamical sublime, 9
  and human reason, 17–18
  on man-made works, 10
  mathematical sublime, 9
  on relationship between sublime and work of art, 6
  by romantic imagination, 3
  and terrible catastrophes, 15
  and work of art, 13
Keyhole Nebula, 69
King Kong. See O'Brien, Willis
Klee, Paul, 13
Knox, Israel, 107
Koch, Christoph, 51
Koko the Clown. See Fleischer, Max
Krebs cycle, 40
Kristeva, Julia, 83
  Révolution du langage poétique, 83
Krushchev, Nikita, 162

La condition postmoderne: rapport sur le savoir. See Lyotard, Jean-François

Lacoue-Labarthe, Philippe, 87
language, 31
  affects as source of, 32–34
  feelings and, 33
  language instinct, 32, 33
Levay, Zolt, 65, 70
libido, 30
Librett, Jeffrey, 77
  Of the Sublime: Presence in Question, 77
Little Nemo. See McCay, Winsor
livingness, 39
Longinus, 45, 85
  Peri Hypsous (On Height), 94
Lorenz, Konrad, 93
lowering of sublime, 52
Lynch, Michael, 60
Lyotard, Jean-François, 12, 14, 77, 79, 80, 95
  La condition postmoderne: rapport sur le savoir, 12

mad scientists, 131
mammals, 27–28
mapping
  body, 37
  viscera, 37, 38
Martin, John, 7
Marx, Leo, 130
material monism, 45
material sublime, 45
maternal devotion, 27–28
mathematical sublime, 9, 102
McCay, Winsor, 132
  Gertie the Dinosaur, 128
  Little Nemo, 132, 138
median sublime, 154–55
Méliès, Georges, 137

mental life
  affective foundations of, 22–27
  dual-aspect monism view of, 23
mental movement, 99
Metropolis, 128, 131, 145
Michelangelo Buonarroti, 47, 141–42, 158
Mickey Mouse. *See* Disney, works by
microscopic perspective, sublime aspect, 152
Milton, John, 47
mind, 23–25
  higher regions of, 40–41
  passions of, 27–29
mind-medicines, 39
  search for, 29–30
modern painting, 78
molecules, 152
Mondrian, Piet, 80
monitoring station, 100
Morris, Desmond, 93
  Naked Ape, The, 93–4
  Pocket Guide to Manwatching, The, 93
Müller, Karlheinz, 103
Mulvey, Laura, 135
music
  and language, 32–34
  and skin orgasm induction, 34–36
My Fair Lady, 142
mysterium tremendum, 109

Naked Ape, The. *See* Morris, Desmond
Nancy, Jean-Luc, 78
NASA, 57, 64
National Socialist regime, Germany, 9
nature, 47–48

nature's terribilità, 45
negative sublime, 14
neocortical expansion, 26
neologisms, 49
neural activity
  and emotional affects, 23
neural soul, 36
neural systems, 28
neuroscience and sublime in art and science, 91–105
Newman, Barnett, 12, 79
  Adam, 12
  Vir Heroicus Sublimis, 12
Newtonian models, 12, 78, 118
nineteenth-century landscape painting, sublime impulse in, 7
nineteenth-century romantic discourse of sublime, 82
Noll, Keith, 64, 71
nonconscious perception, 53
nonconscious sublime, 43–55
norepinephrine, 40

objective explanations, 51
O'Brien, Willis, 130
  King Kong, 131, 145
Observations on the Feeling of the Beautiful and the Sublime, 94. *See* Kant, Immanuel
Of the Sublime: Presence in Question. *See* Librett, Jeffrey
Onslow-Ford, Gordon, 80
On the Origin of Species, 156. *See* Darwin, Charles
opioids, 28–29, 30
orientation, 61, 62
Otto, Rudolf, 85–86
  Idea of the Holy, The, 85

Out of the Inkwell. See Fleischer, Max
oxytocin activity, in brain, 28

Pagels, Heinz, 24
painful loneliness, 29
painting, 12, 48, 76, 78
Parkinsonian stillness, 31
Peri Hypsous (On Height). See Longinus
Philosophical Enquiry into the Origin of Our Ideas of the Sublime and Beautiful, A. See Burke, Edmund
physical contact, 46
physiological sublime, 97
Pico della Mirandola, Giovanni, 155
pictures, prettiness of, 61, 62, 65–66
picturesque and prettiness, 59
Pinocchio. See Disney, works by
Platonic mathematical idealism, 124
Pocket Guide to Manwatching, The. See Morris, Desmond
poor sublime, 86–89
Popper, Karl, 159
positive sublime, 14
post-Kantian image sublimity, 79
postmodern sublime, 12–13, 14, 75, 78
postquantum mechanics, 122
Powell, Colin, 17
prehension, 49
prettiness, 57
  as an aesthetic concept, 58–59
  and color images, 60
  and femininity, 59
  Hubble images, 57–58
  of pictures, 61, 62, 65–66
  and picturesque, 59
prolactin, 27

Prometheus Unbound (Percy Bysshe Schelley), 52
proto-music, 33, 35
Pushkin, Aleksandr, 7
  Boris Godunov, 7
Pygmalionesque imagination, 136

quantum romanticism, 106–25
  aesthetics manifests, 111
  mysterium tremendum, 109
  regulative metaphors, 110
  spiritual rehabilitation, 109

Räuber, Die. See Schiller, Friedrich
reason, human, 17–18
religious concept
  in art criticism and history, 86
  and sublime, 84–86
Révolution du langage poétique. See Kristeva, Julia
Rilke, Rainer Maria, 162
Riskin, Jessica, 141
Romantic discourse of sublime, 82
Romanticism, quantum, 106–25
Romantic Sublime: Studies in the Structure and Psychology of Transcendence, The. See Weiskel, Thomas
Rorty, Richard, 87
Rothko, Mark, 12
ruthless reductionism, 24

Sacks, Oliver, 152
  Uncle Tungsten, 152
Sarfatti, Jack, 120
Schelling, Friedrich, 51, 117, 118
Schiller, Friedrich Wilhelm, 7
  Räuber, Die, 7

science, sublime in, 75–77, 149
scientific communication, 162
seeking system, 30–31, 40
self, 37
self-consciousness, 46
selfhood, 49
September 11, 2001 events, 14–15
Shapiro, Gary, 9
Shaw, Philip, 17
Simmel, Georg, 47
Six Stories from the End of Representation. See Elkins, James
skin orgasms, musically induced, 34–36
Snow White and the Seven Dwarves. See Disney, works by
social emotions
  and artistic endeavors, 33
solipsism, 51
somatic markers, 53
Sorcerer's Apprentice, The. See Disney, works by
souls
  neurobiology of, 36–38
  of human and animal, 38–41
Space Telescope Science Institute (STScI), 57, 64
Speer, Albert, 10
  "Große Halle," 10
spiritual rehabilitation, 109
Stafford, Barbara Maria, 43
  Echo Objects: The Cognitive Work of Images, 54
Steiner, George, 78
Stengers, Isabelle, 51
Stewart, Susan, 134
Sticky Sublime, The. See Beckley, Bill
strange loop of selfhood, 52

Sublime, The. See Tsang, Lap-chuen
sublime object, 4–5
superject, 49
supersensible capacities, 5–6

Taylor, Mark C., 77
  Disfiguring: Art, Architecture, Religion, 77
Taylor, Nicholas, 8
terror
  aesthetics of, 15
  sublime as, 14, 15–16
text vs. visual art, 6
Thagaard, Paul, 51, 53
theology, 108
Tinbergen, Nikko, 93
Todorov, Tzvetan, 83, 129
transcendental sublime, 12, 97
Treatise of Human Nature. See Hume, David
Tsang, Lap-chuen, 79, 81
  Sublime, The, 79
Tsvetaeva, Marina, 162
Turner, J. M. W., 7
2001: A Space Odyssey, 128

Über das Geistige in der Kunst. See Kandinsky, Wassily
Uncle Tungsten. See Sacks, Oliver

ventral stream, 98, 99
Victorian cities, Britain, 8
Vidler, Anthony, 129, 134
Vir Heroicus Sublimis. See Newman, Barnett
visceral brain regions, 38
Vischer, Friedrich Theodor, 8, 9
visual art, sublime in, 11, 13, 77–78

Weiskel, Thomas, 84
  Romantic Sublime: Studies in the Structure and Psychology of Transcendence, The, 84
Wenders, Wim, 163
  Wings of Desire, 163
western metaphysics, 80
Whale, James, 130
Wheeler, John, 111, 120, 125
Whirlpool Galaxy, 62, 63, 70
Whitehead, Alfred North, 44, 48, 49, 54
Wiles, Andrew, 160
Winckelmann, rule of beauty, 46
Wings of Desire. *See* Wenders, Wim
Wood, Gaby, 136
Woodward, R. B., 160–61
work of art and sublime, relationship between, 6

Zeitgeist, 161
Zeki, Semir, 93, 99, 100
  Inner Vision, 93, 99
Žižek, Slavoj, 5